Sustainable Wildlife Tourism

Abel Ansporthy Mamboleo

Sustainable Wildlife Tourism

Springer

Abel Ansporthy Mamboleo
Department of Tourism and Hospitality Management
The Saint Augustine University of Tanzania
Mwanza, Tanzania

ISBN 978-3-031-97794-7 ISBN 978-3-031-97795-4 (eBook)
https://doi.org/10.1007/978-3-031-97795-4

This Springer imprint is published by the registered company Springer Nature Switzerland AG
The registered company address is: Gewerbestrasse 11, 6330 Cham, Switzerland

I dedicate this book to my grandmothers, Tupokigwe Ngoka and Gomokaga Kijumile; my father, Ansporthy Mamboleo Mwakaleja; my mother, Getruda Kituli Mwakubombaki; my uncles, Abdallah Kisinga Mwakaleja, Simon, and Amani; my wife, Dr. Mida Abel Msamba, whose unwavering companionship and warmth I could never repay; my daughters, Judycia and Getrucia; and my son, Nelson, who consistently loved and supported me, even when I was on the verge of giving up writing this book. Your spiritual, financial, and health support made all of this possible for me.

Preface

Wildlife tourism comprises watching, enjoying, and learning about wild flora, animals, and the natural environment in protected and unprotected wildlife landscapes. Besides the enjoyment, relaxation, and education, it also provides myriad advantages to wildlife tourism stakeholders. Wildlife tourism occurs in protected and unprotected wildlife habitats, usually owned publicly, communally, and privately.

This underlines that wildlife tourism may take place in areas of all types, whether in protected or unprotected regions or open areas where people continually interact with animals. Protected landscapes are formally governed by national or international legislation, but they are additionally regulated by customary norms.

As a result, wildlife tourism destinations include national parks, forest reserves, wilderness areas, and natural monuments, among other things. All these are at the heart of wildlife tourism, where key tourist activities occur. Such a region has unique, well-protected animal resources that attract wildlife visitors from all over the world.

The most significant aspects of wildlife tourism are the availability, sufficiency, attractiveness, and accessibility of wildlife resources. However, wildlife tourism does not just rely on protected areas; it also depends on the status of unprotected areas, including open areas.

According to studies, several wildlife species, such as elephants and zebras, spend more than six months outside Tanzania's protected areas every year. Surprisingly, 80% of African elephants graze and reproduce outside the protected habitat.

That is, unprotected areas have equally attractive wildlife species to attract wildlife visitors. However, the availability of suitable animal resources in such areas is not the only criterion for upgrading an unprotected area to a protected one.

Wildlife tourism goes beyond providing enjoyment, relaxation, and education to wildlife visitors; it also benefits governments, protected areas, local communities, tour operators, and hotels. It provides a wide range of practical and intangible benefits to wildlife stakeholders.

Wildlife tourism provides social, cultural, and environmental benefits to stakeholders. Wildlife tourism provides job opportunities and builds social facilities such as hospitals, schools, and local markets. Wildlife tourism contributes to the destination's economic growth by creating revenues and profits in foreign currency. Environmentally, it sensitizes wildlife tourism stakeholders to directly and indirectly assist in environmental protection via contributions, research, and education.

However, uncontrolled wildlife tourism destroys the ecology, community, and economy. Conventional wildlife tourism disrupts the natural ecosystem through disease transmission, changes in animal behavior, wildlife population dynamics, species endangerment, deforestation, wildfires, land degradation, and water and air pollution. Also, wildlife tourism impacts communities by escalating social crimes, relocating local people, causing unemployment, creating a shortage of local resources, over-commodifying culture, and mismanaging wildlife resources.

Additionally, conventional wildlife tourism seriously challenges the host communities, conservationists, tourists, and other stakeholders by failing to effectively conserve those places with high biodiversity, to involve all key wildlife tourism stakeholders in wildlife tourism and conservation, and to increase wildlife tourism revenues and foreign earnings for destinations. Wildlife tourism also fails to avoid its detrimental effects and enhance its positive repercussions. It needs to encourage local investors in wildlife tourism and undertake meaningful wildlife tourism research.

This book advocates transitioning from a conventional to a sustainable wildlife tourism paradigm via the adoption of sustainable development principles as one of the key tactics for reducing or eliminating the unfavorable effects and maximizing the positive ones. Appropriating environmental, social, and economic sustainability concepts may overcome most issues related to wildlife tourism.

Despite our in-depth understanding of wildlife tourism's positive and negative aspects, our understanding of its integration with the concept of sustainable tourism still needs to be explored. This book provides a detailed conceptual framework and procedure for integrating the different concepts and approaches to sustainable tourism, focusing on the frequently overlooked link between conventional and sustainable tourism. It emphasizes the advancement of sustainable wildlife tourism at both the attraction and destination levels.

Based on the theoretical implications, the book adds to the existing knowledge of sustainable tourism development by providing a framework for a better understanding and integrating conventional and responsible medical tourism. It also examines the transition from a conventional to a sustainable wildlife tourism paradigm. It supports the argument that implementing sustainable development principles is one of the key solutions for reducing or eliminating related impacts from wildlife tourism.

Wildlife tourism players need to demonstrate personal honesty and dedication to sustainable wildlife tourism. Many wildlife tourism stakeholders have adopted sustainability principles conceptually, not practically. Even newly developing wildlife tourism entities and destinations preach about the sustainability of wildlife tourism, but they still use the same old tactics. However, inadequate legal-policy frameworks, reluctance, and ignorance are among the most common reasons for stakeholders' neglect of sustainability principles.

Moreover, this book addresses the unsustainability problem of wildlife tourism by suggesting the adoption of guiding principles of responsible tourism, which are bottom-line strategies for implementing sustainable wildlife tourism. However, the adoption of sustainable tourism comes with certain expenses and requirements. It may necessitate that the destinations undergo fundamental changes in their existing legal and policy frameworks.

Furthermore, this book has practical and theoretical implications for academics, wildlife and tourism professionals, students, policymakers, and wildlife tourists with or without sufficient background in wildlife tourism. It introduces readers to the basic concepts of wildlife tourism, responsible tourism, and sustainable tourism and explores traditional wildlife tourism's various dimensions, impacts, and challenges.

This book is essential for wildlife tourism stakeholders, including tour companies, governments, international and local organizations, students, protected area managers, hotel operators, travel operators, and tour guides. Wildlife travel operators can use this book to align their operations with sustainability practices, such as engaging in community and social responsibilities and committing to environmental responsibilities.

They can do this by directly supporting responsible environmental practices or educating tourists and other stakeholders on responsible wildlife tourism. Therefore, it provides technical guidance for wildlife tourism managers to improve their practices regarding wildlife conservation and tourism operations.

Nature conservationists and tourism practitioners may find this book pertinent for advancing their knowledge, practices, and skills in balancing tourism operations and wildlife conservation. Balancing wildlife conservation and tourism operations has been difficult, especially in the developing world, where some protected areas mainly depend on tourism operations to fund wildlife conservation activities.

The book provides practical insights that wildlife and tourism managers may use to balance nature conservation and operations. Conservationists and tourism managers may use the sustainability model, framework, and cases provided in this book when balancing the needs of two operations.

It is helpful to policymakers and development agencies that oversee global wildlife and tourism. The text gives in-depth guidance on sustainable wildlife tourism and offers models for legal and policy frameworks. Policymakers may use the models in wildlife tourism policies and agreements at all levels.

Local and international organizations, including UNWTO, SIDA, WWF, AWF, GEF (Global Environment Facility), and USAID, may find this book valuable when designing and implementing sustainability programs that promote environmental, social, cultural, and economic responsibilities. The cited case studies in this book demonstrate how the effective tourism policies of Bhutan and Singapore have significantly shaped and enhanced the sustainability visions of the countries.

Moreover, this book suits graduate and undergraduate students studying tourism, environmental management and policy, hospitality operations, and nature conservation. It provides students with sustainability concepts and practices, particularly integrating the three pillars of sustainability with tourism operations. Furthermore, general readers and wildlife enthusiasts may find this book helpful as it explains various issues that directly or indirectly affect them, including their interactions with wildlife and travel to wildlife tourism destinations.

The information offers valuable guidance to policymakers and development agencies responsible for overseeing wildlife and tourism resources on a global and international scale, as it presents detailed models of sustainable wildlife tourism for legal and policy frameworks. With such insights, policymakers may integrate the models into national, regional, and global wildlife tourism policies, agreements, and enforcement strategies.

Mwanza, Tanzania Abel Ansporthy Mamboleo

Acknowledgments

This book may not have been written without the encouragement and ideas of the thousands of students at the undergraduate level about sustainable wildlife utilization, wildlife tourism, sustainable tourism, and ecotourism courses that Abel Ansporthy Mamboleo taught at the Saint Augustine University of Tanzania (SAUT) and many technical colleges in Mwanza City for over 12 years.

I especially have to extend my thanks to my awesome wife, Dr. Mida Abel Msamba, for reading the initial drafts, providing constructive criticism on the arrangement of chapters, suggesting the book's contents, and even selecting the publisher.

Also, my previous colleagues and PhD research supervisors were helpful in the authoring and development of this book: Dr. Crile Doscher, Prof. Adrian Patterson, Sue Bowie, Professor Fransesc, and Dr. Yaa Akwaaba. Abel has also expressed gratitude to the students in his wildlife-related courses and those who attended his Sustainable Tourism and Sustainable Wildlife Utilisation seminars at SAUT, Tanzania, and Lincoln University, New Zealand.

Abel's teaching classes and PhD program have been the workshop for working out ideas, integrating new ones, and sharing his experiences as an author and researcher. In addition, Abel wishes to thank his team in the Department of Tourism and Hospitality Management at SAUT, who have helped design this book's content throughout the years.

In addition, I am thankful for the constructive comments made by anonymous reviewers Aaron Schiller and the editor since I could not have produced this work without my reviewers' great support, analytical comments, and encouragement. I especially want to thank my wonderful family for their encouragement and support throughout my writing.

Without the experiences and support of my previous employers and colleagues at the Ministry of Natural Resources and Tourism, Wildlife Division, Mvomero District Council, the then-Wami Mbiki Wildlife Management Area, and DOMMA Wildlife Area, this book would not exist.

You gave me a special opportunity to practice what I learned at the universities and earn the world-class experience that I am proud of today. Thank you, Mr. Kunambi, Mr. Kawili, Mr. Fortunatus, Mr. Macha, Mr. Mwaikenda, Mr. Byelushengo, Mr. Mushi, Mr. Mkumbo, and Mr. Chambile.

Chapters Outline

This book is divided into three key parts and eight chapters. Part I includes Chaps. 1 and 2, which highlight game species, wildlife species, and protected areas as tourist resources and places where wildlife tourism occurs. Chapters 3 and 4 belong to Part II, which covers the various types of wildlife uses and environmental and socio-economic effects of wildlife tourism. Part III exclusively emphasizes the need for adoption and integration of responsible and sustainable development concepts into wildlife tourism as a means of eliminating the adverse impacts of wildlife tourism.

Chapter 1: Wildlife and Game Species

In this chapter, the author starts by introducing the reader to important wildlife and game terminologies, as widely and frequently applied in wildlife tourism. Wildlife and game species are fully discussed, in this chapter, allowing readers to have a common understanding of the subjects throughout this book. Also, the chapter presents the readers with traditional wildlife tourism. Thus, this chapter should assist the leaders in comprehending the distinction between game and wildlife species, and the significance of adopting sustainable wildlife tourism.

Chapter 2: Wildlife Tourism and Protected Areas

It was necessary to thoroughly analyze protected areas for readers to understand the areas where wildlife tourism takes place. This chapter explains a variety of wildlife protected areas, including local and international categorizations, how they are formed, and how they are connected to sustainable wildlife tourism. This chapter should assist readers in critically examining the applicability of protected and unprotected environments to wildlife tourism growth.

Chapter 3: Wildlife Tourism Utilizations

Wildlife tourism is at the center of wildlife tourism. Wildlife tourism encompasses both consumptive and non-consumptive uses of wildlife resources. This chapter describes several forms of wildlife tourism activities, the way and locations each activity takes, the cautions, and required skills, licenses, and permissions for each activity. This chapter helps readers understand several forms of wildlife tourism activities taking place in protected and unprotected landscapes. Further, it helps readers evaluate and arrange efficiently their wildlife-based travels to various wildlife tourism areas.

Chapter 4: Impacts of Wildlife Tourism

It is necessary to have an in-depth understanding of the benefits of wildlife tourism, from household to destination level. Thus, this chapter examines solely the socio-economic and environmental advantages of wildlife tourism. It starts by introducing the socio-economic and environmental advantages of wildlife tourism. The book then uncovers various instances of wildlife tourism's contributions to socio-cultural, economic, and environmental conservation.

It is also useful to understand the opposite side of wildlife tourism. Any unmanaged wildlife tourist activity may have various social, cultural, economic, and environmental implications. This chapter starts by identifying and addressing probable harmful social, cultural, economic, and environmental implications, followed by an analysis of issues confronting wildlife tourism.

Chapter 5: Sustainable Wildlife Tourism

At the beginning of this book, the author identifies the shift from conventional to responsible types of wildlife tourism as a key subject. In this chapter, you discover the ideas and principles of sustainable development, and how such principles may enhance the quality of wildlife tourism. Furthermore, the readers may learn about options for transitioning from conventional to sustainable wildlife tourism.

Chapter 6: Planning for Tourism Sustainability

Participative planning, innovations, tourist satisfaction, and safety are all elements of sustainable wildlife tourism. In such a situation, sustainable wildlife tourism is a two-way process that involves the satisfaction of wildlife tourism stakeholders, and the commitment of wildlife stakeholders to the environmental, social, cultural, and economic challenges. This chapter highlights to the readers on key elements of human dimensions, including satisfaction, expectations, motivation, gratitude, and support of wildlife visitors.

Chapter 7: Responsible Wildlife Tourism

This chapter presents the roles of wildlife tourism stakeholders in implementing the goals of sustainable tourism while reflecting on the satisfaction and ethics of various players of sustainable wildlife tourism.

Chapter 8: Challenges in Sustainable Wildlife Tourism

This chapter discusses the technical and policy challenges facing the execution of goals of sustainable tourism from individual to institutional levels. Moreover, this chapter pinpoints the challenges of existing sustainable tourism certification schemes. It provides the cases or lessons from two responsible and sustainable destinations that demonstrate practically tourism responsibility and sustainability.

Contents

About the Author

Abel Ansporthy Mamboleo is a Senior Lecturer in the Department of Tourism and Hospitality Management at Saint Augustine University of Tanzania. He has authored numerous articles on wildlife tourism, conservation, and computational modeling.

While at the Saint Augustine University of Tanzania, he served as the head of the Department of Tourism and Hospitality Management for five years (2020 to 2025) and is one of the founders of the *Eastern African Journal of Hospitality, Leisure, and Tourism*.

He worked as the District Wildlife Officer for over 6 years in the Morogoro region, where he practically managed to jointly work with several local and internal organizations as well as community-based wildlife conservation organizations.

He attained his first and second degrees in Wildlife Management at Sokoine University of Agriculture, the public university located in Morogoro, Tanzania. He did his Ph.D. in Environmental Modeling and Simulation, specializing in human-elephant interactions in the communities adjacent to Serengeti National Park, at Lincoln University, Christchurch, New Zealand.

Abbreviations

AIDS	Acquired Immunodeficiency Syndrome
AWF	African Wildlife Fund
BEC	Bureau of Economic Analyst
CAMPFIRE	Communal Areas Management Programme for Indigenous Resources
CBD	Convention on Biological Diversity
CBNRM	Community-Based Natural Resources Management
CITES	Convention on International Trade of Endangered Species
COVID-19	Corona Virus Disease of 2019
GDP	Gross Domestic Product
GEF	Global Environmental Facility
HIV	Human Immunodeficiency Virus
IFAD	International Fund for Agricultural Development
ILO	International Labor Organization
IUCN	International Union for Conservation of Nature
KWS	Kenya Wildlife Services
MDPR	Minimum Daily Package Fee
MMNP	Mahale Mountains National Park
NCAA	Ngorongoro Conservation Area Authority
PAs	Protected Areas
RAMSAR	Iranian City where the Convention of International Wetlands occurred
SANPARKS	South Africa National Parks
SAUT	Saint Augustine University of Tanzania
SDF	Sustainable Development Fee
SIDA	Swedish International Development Cooperation Agency
SWOT	Strengths, Weaknesses, Opportunities and Threats
TANAPA	Tanzania National Parks
TAWA	Tanzania Wildlife Management Authority
UNEP	United Nations Environment Programmes
UNESCO	United Nations Scientific and Cultural Organization
UNWTO	United Nations World Tourism Organization

USAID	United States Agency for International Development
WCMC	World Conservation Monitoring Centre
WHO	World Health Organization
WMAs	Wildlife Management Areas
WTTC	World Travel and Tourism Council
WWF	World Wildlife Fund for Nature

Part I
Wildlife Tourism and Conservation Areas

Chapter 1
Wildlife and Game Species

Abstract This chapter analyses the fundamental concepts of wildlife and game species, emphasizing their significance in wildlife tourism. Wildlife includes naturally occurring, free-ranging invertebrates and vertebrates, namely animals, fungi, bacteria, and protozoa inhabiting their natural habitats with minimal human influence. Wildlife species are crucial for preserving and maintaining the dynamic equilibrium of their natural habitats by supporting vital ecological processes, including mating, migration, feeding, and symbiosis. Captive individuals of wildlife species in zoological gardens or zoos are rarely considered as part of wildlife, despite lions, elephants, and zebras being the products of free-ranging species. In both protected and unprotected areas, wildlife species autonomously seek ecological resources, including water, food, and shelter, without reliance on humans. While fences and other ecological changes are crucial for species protection, they can sometimes conflict with natural processes and create substantial issues for wildlife habitats. This chapter underscores the essential function of wildlife species in ecosystems, contributing to ecological resilience and wildlife tourism development. It discusses the role of wild animals and plants in the development of wildlife tourism, using Kitulo Plateau National Park as an illustration of a tropical protected area established for the management and conservation of wild flora. The national park is renowned for its seasonal flowering of natural flora, drawing hundreds of nature enthusiasts interested in wildflowers. Such wildlife visitors directly and indirectly help in the preservation of the park for current and future generations. The game species refers to wildlife that may be legally hunted for trophies or valuable components. Various species can be hunted either seasonally or throughout the year for recreational, traditional, and even meat purposes. In short, the game is a subset of wildlife species, and its vocabulary is largely applied to wild animals and not plant life. In many developing countries, wildlife hunting was introduced by the colonial administration during the colonial era. However, ethical and conservation considerations have prevented the killing of some species and terminated entirely tourist hunting, especially in certain wildlife destinations. Various animals, including elephants and black rhinoceroses, are exclusively prohibited by various international and national regulations from hunting business on some hunting grounds. Also, in several nations, such as Kenya, hunting is not a component of wildlife tourism. It is crucial

A. A. Mamboleo, *Sustainable Wildlife Tourism*,
https://doi.org/10.1007/978-3-031-97795-4_1

to emphasize that effective biodiversity conservation relies on the balance between social, cultural, economic, and ecological sustainability.

Keywords Wildlife · Biodiversity · Conservation · Game species · Sustainability · Ecosystem · Habitat · Wild animals

During their visits to wildlife attractions, most wildlife visitors are delighted to encounter magnificent animal species in their associated natural or constrained areas. The various behaviors displayed by wildlife resources, such as sounds, colors, activities, mating, and interaction behaviors, make wildlife tourists excited and even intrigued, desiring to learn more about the observed species.

In that aspect, the animals in their natural habitats become even more fascinating because, in reality, they display their normal activities without or with the least interference from human beings. This is why many visitors go on lengthy journeys from their homes to faraway wildlife destinations to view animal resources. Some wildlife tourists save money for years before going to those attractions with magnificent wildlife resources.

In terms of behaviors, sightseeing, game-watching, or game-driving is one of the most desired methods of wildlife tourism. Visitors drive or walk about to enjoy the varied behaviors of wild animals with or without the help of tour guides. Tourists generally witness animals' spectacular anatomical and behavioral qualities during game drives.

It is practical to declare that game driving is not the only type of wildlife tourism in protected and unprotected wildlife habitats. Other enormous types, like hunting and fishing, are quite widespread, particularly in the developed world. Despite its magnetism and beauty, wildlife is a resource whose exploitation must be monitored and cross-checked regularly.

Wildlife may be unlawfully and carelessly consumed as a resource, become scarce, and occasionally face serious extinction. It demands thoughtful usage and frequent conservation measures, not only for the coming generations but also for the present generation.

Until now, inattention to preserving wildlife resources has taught the world many painful lessons. The extinction of certain attractive species of animals is one of the worst lessons taught by neglectful behavior. This is why this book integrated wildlife tourism with the concepts of sustainability, helping readers understand the necessity of wisely utilizing environmental resources.

With such notions, tourists, tour operators, tour guides, and conservation authorities must exploit natural resources wisely for the benefit of current and future generations. Responsible utilization of wildlife resources may improve their lifespan and produce greater delight for present and future generations.

In other words, each wildlife tourism stakeholder should be responsible before, during, and after visiting wildlife habitats. The stakeholders may actively participate in safeguarding the resources by caring for the communities that live near wildlife areas and contributing to the conservation of wildlife resources through donations, research, and initiatives.

For similar reasons, wildlife tourists and tour operators should protect wildlife resources by not feeding the animals in their natural refuge and not being involved in wildlife crimes such as poaching, causing wildfires, off-road driving, and introducing invasive alien species to natural regions.

Their engagement in the crimes above may disturb the animals' social structure and interfere with the natural environment, thus leading to local and global extinctions of wild flora and animals. However, sustainable wildlife tourism encourages tourists and tour operators to participate in developing wildlife resources by directly or indirectly supporting conservation programs by contributing to and participating in specific activities such as the uprooting of invasive species.

Sustainable wildlife tourism encourages participants to excel in moral and ethical tourism by promoting community livelihoods. Studies indicate that communities near wildlife regions where wildlife tourism is prevalent frequently confront undesirable socio-economic implications as outcomes of unsustainable tourism and wildlife.

Wildlife tourism in areas with insufficient security measures is harmful since it can quickly convert such wildlife destinations into criminal hotspots. In some destinations, there is a probability of increasing prostitution, drug addiction, poaching, unemployment rates, young people not being active in production sectors, and poaching occurrences, among others.

Furthermore, since wildlife is the main attraction of sustainable wildlife tourism, conservation authorities may intensify their conservation efforts, increasing the number and variety of wildlife species. On the other hand, this may impact the intensification of incidents of human–wildlife interactions in the adjacent communities.

As a result, local people may have restricted their involvement in agricultural operations to minimize agricultural losses from problem animals. Since agriculture is the backbone of the rural economy, their abstention from agriculture would render local people impoverished and beggars.

Introduction to Wildlife

It is essential to have a common understanding of wildlife terminology before proceeding with further illustration and description of sustainable wildlife tourism. Conservationists have attempted to describe the meaning of wildlife in various ways, using various terminologies while citing several examples.

Since this book is solely about sustainable wildlife tourism and not purely conservation, the author attempts to elucidate the meaning of wildlife in a way that most readers can understand without contradicting its actual meaning and essence.

The term wildlife derives from two distinct meanings: untamed and living. It alternatively signifies existence in the wilderness or anything existing in the wilderness. In the wilderness, many living creatures exist, ranging from vertebrates to invertebrates and vegetation to animals. In this context, life implies something that can independently move, consume, breathe, and communicate, but it should reside in the wilderness.

Based on the description above, life should exist in the wilderness, its natural environment. African buffalo, *Syncerus caffer caffer*, in the zoological garden or zoo is not wildlife because the zoological garden or zoo is not a naturally occurring environment for the animal, as any animal in the zoological garden can be as good as any other livestock, such as cows or lambs, among others.

It is safe to clarify that inhabiting natural areas is not the only characteristic that makes an animal wildlife. This brings us back to the definition of wildlife in the previous paragraph, which characterizes wildlife as vertebrates and invertebrates, the next characteristic of wildlife to discuss.

The wilderness is the habitat of many varieties of organisms, including animals with and without vertebral columns. Let us simplify the terminologies: When naturally occurring in the natural environment, all organisms in the kingdom Animalia, including grasshoppers, fish, reptiles, birds, bacteria, fungi, viruses, and protozoa, are equally regarded as wildlife.

The same applies to wild plants, including acacia, baobab, and various botanicals, which can be classified as wildlife in the wilderness. The definition of wildlife seems to contradict what people are used to understanding about wildlife. For example, the definition highlights that the African buffalo in the cage is not a wildlife species. This is true because the domestication of African buffalo disqualifies the animal from being a wildlife species.

Though most people identify the wilderness of any animal by name, not context, according to them, lions and African elephants will always be wildlife, whether in the zoo or wilderness. That also raises another issue: Does introducing sheep, goats, or cattle into the national park make them wildlife? Some people may still perceive the introduced species as a domestic animal.

Though the answer depends on the reasons behind the introduction, in some protected areas, livestock keepers illegally encroach on the parks for grazing in the dry season. This does not make the cows both domestic animals and vermin or pests, but if the cow is legally introduced for scientific or conservation purposes or any other legal purpose, it may qualify as wildlife, though other factors will have to be considered.

Additionally, confusion may arise about the clarification on wilderness or natural environment. What does it mean by wilderness or a natural environment? Does it only mean national parks or protected areas? In other words, for animals or vegetation to qualify as biodiversity, should they only occur in national parks or conservation areas? What about the creatures that cross the park boundaries and occupy unprotected or human landscapes for more than 6 months? When they are in human territory, should they still be considered wildlife?

In this context, the natural landscape or wilderness does not only imply protected areas; studies show that most protected areas were incorrectly established, so most animals spend months outside the designated areas. Also, the absence of fundamental resources in the protected landscape sometimes necessitates the animals migrating to other areas to locate the resources.

More importantly, the need for more breeding partners and reproductive spaces forces the animals to migrate outside the protected areas. In brief, an animal can be considered wildlife even outside of the protected areas as long as it has the basic features and behaviors of wildlife.

Furthermore, apart from the mentioned characteristics, the following essential feature of fauna is caging-free or free-ranging. Unlike domestic animals that are always constrained, wildlife should be free-ranging. In other words, wildlife should be free-ranging by not existing in cages or being tamed like livestock. There should be no or limited human interference on where, when, or what to feed, drink, and mate. Wildlife in the natural environment lets nature control most of the processes.

Wildlife decides where to migrate to another area, consume, and drink differently from animals in the zoo or livestock, whereas humans decide everything for animals. Confined animals always get frequent treatments. Also, to improve their reproductive quality and productivity, humans usually choose the best breeding partners, while in the wilderness, wildlife does that itself.

Consequently, wildlife usually migrates from one part of the ecosystem to another, searching for potential and relevant breeding mates or ecological resources. The wildlife then goes outside the protected areas to locate nutritious food to improve their reproductive success. Other species, like African elephants, chimpanzees, and baboons, may cure themselves by consuming medicinal plants or licking rocks.

However, some protected areas construct fences around their borders to control the movement of animals after perpetrating endless socio-economic problems in the neighboring communities. This situation does not disqualify the animals from being wildlife. However, it can be dangerous as it may cause inbreeding and over-predation, resulting in the extinction of some species that cannot cross the fence to prevent mating with their closest relatives.

In many cases, the fences tend to control only crawling and walking, not climbing or flying species like birds. The fencing may also be appropriate for small and medium-sized protected areas, as large protected areas like Serengeti National Park are geographically, ecologically, and financially impracticable. In that situation, only a part of the protected area may be fenced, or the problem species may be fenced within the protected areas.

Furthermore, certain circumstances contradict the concepts of cage-free and free-ranging wildlife, whereby some protected areas have caged certain animals within the protected landscape. The categorization of such animals into wildlife can still be questionable because, before commenting on that circumstance, it is crucial to understand the reasons behind such decisions. Also, caging or free-ranging are not the only criteria that make animals wildlife; therefore, it is necessary to understand the context in which the animal was confined.

Some conservation authorities do not treat sick animals, but during disease outbreaks, the diagnosed animals are isolated and confined for natural recovery;

similarly, some animals may be caged for scientific investigations, and others may be caged for tourism purposes.

In the context of tourism, which is the primary talking point of this book, some protected areas, particularly those located near large cities, tend to operate like management. Such wildlife areas tend to have a variety of untamed animal species and vegetation to entice domestic travelers.

By doing so, the protected areas sensitize domestic tourists, encouraging them to visit other protected areas. In such cases, protected areas translocate certain attractive animals from other protected areas to increase the park's diversity and attractiveness. As a result, the protected area cannot meet the habitat and social requirements of the translocated animals, as many translocations need to comply with ecological directives.

In that case, caging is the only option for avoiding over-predation, human–wildlife conflicts, and exploitation. Since caging necessitates the animal's sustenance and restriction of their movements, it inherently disqualifies all caged animals from being fauna.

When I introduced the concept of wildlife, I explained that it entails wild animals or wild plants. Regrettably, most of my discussions have relied on wild animals, with minimal narration on wild flora.

Let me describe the wild plants by posing this question: "Can wildlife tourism take place in protected areas that do not have wild animals but only wild plants?" Based on the definition of wildlife, the answer to this query is undoubtedly yes because untamed flora is wildlife.

Many areas are well known for the diversity and attractiveness of their vegetation, which attracts several visitors. Kitulo Plateau National Park, also known as the Serengeti of Flowers, exemplifies this circumstance. In this context, the behavior of plants, which is the diversity and synchronized flowering of plants in certain months of the year, attracts visitors to venerate them. Various untamed flora's physical and physiological features are always essential assets for wildlife tourism.

Some protected areas have the longest trees (Lau Catchment Forest Reserve), the oldest trees (Baobab trees, *Adansonia digitata*, in the Ruaha National Park), and several other marvels to attract visitors. Untamed plants occur in protected and unprotected areas, including forests and wildlife reserves. However, it depends on the national wildlife and tourism policy, and some countries prefer to direct the actual meaning of wildlife.

Additionally, this segment emphasized more about the naturally occurring environment, but regrettably, it is still uncertain how the environment can be a constituent of wildlife. In rudimentary ecology or biology, some of us learnt about the environment as a collection of living and non-living objects. In that classification, fauna is one of the living components of the environment, while rocks, mountains, rivers, and soil are non-living.

In this concept, the primary discussion is about the natural environment, not an artificial environment like a zoological or botanical garden. In summary, the natural environment directly sustains life through the provision of food, water, and shelter to wildlife; think of vegetation as food for African buffalo; think of African buffalo as food for lions. All these are biological components of the natural environment.

Furthermore, baobab trees should be considered a shelter or home for leopards, and acacia trees should be considered as sustenance for herbivores. Furthermore, the boulders, or kopjes, in Serengeti National Park are home to reptiles, and the wetland is home to aquatic species but also serves as water reservoirs for every living creature in the wilderness. As a result, the naturally occurring environment is, in essence, a fauna.

Introduction to Game Species

The previous discussion about the definition of wildlife brings us to the next concept: game species. It is easy to confuse these two concepts; even some conservationists still fail to differentiate them correctly. For example, it is common for some conservation agencies to describe their personnel as game or wildlife officers. However, when you ask, particularly the personnel themselves, what makes the wildlife officer distinct from the game officer, it may result in a scrambled response.

A minor difference exists between wildlife and game species. The game encompasses all species that are legally hunted for a variety of reasons, including trophies and leisure. It is common to find this terminology in destinations where hunting tourism is legally permitted, including Tanzania and South Africa. Chapter 6 provides comprehensive illustrations of hunting tourism activities.

The variation between game and ecological species depends on the fact that game species are mostly vertebrates, not invertebrates and plants. The consumptive values of such species led to the coinage of such a name, as most species are usually pursued for their trophies and leisure.

In the African context, during colonial regimes, the areas with large concentrations of trophied animals were declared game reserves, not wildlife reserves, as the conservation policies of colonial regimes prioritized the conservation of hunted species more than others. That was distinct from modern conservation, where all species receive a similar intervention.

Since then, the nomenclature of some protected areas where hunting tourism still takes place has maintained their names suffixed with the phrases "game reserve" or "game parks." For example, Selous Game Reserve, Maswa Game Reserve, and Rungwa Game Reserve. The same concept extends to private areas owned by private entities where hunting can legitimately take place; the names of the sanctuaries are "game ranch" or "game farmers."

Conservationists also retain the former designation of game officers instead of wildlife officers. It is essential to clarify that designating wildlife attractions as game reserves or conservationists as game officers is neither illegal nor illegitimate. Responsible conservation agencies or governments decide whether to retain or enhance the nomenclature depending on the circumstances.

A detailed difference between game and wildlife species is that wildlife sounds like parent terminology because all game species are part of wildlife. In other words, all game species are wildlife species, but not all wildlife species are game species. However, this statement should be cautiously made by remembering that we have just learned about wildlife, which is free-ranging and confined.

Again, we have recently illustrated that game species can also be raised in private game ranches and farms where intensive care is the owner's priority for the increment of products and value addition. In such a context, game species fail to be wildlife species; instead, they become domesticated animals. It is important to note that a trophy can be any portion of the animal that the hunters pursue, including antlers, skulls, skins, or hooves, among other things.

Not all vertebrates are game species because not all are pursued or wildlife species. Regarding national and international laws, not all vertebrates can be hunted in particular countries. With that understanding, some vertebrates, such as black rhinoceros, have been excluded from the hunting quota and are considered just wildlife.

The fundamental reasons for excluding certain species vary from ecological to socioeconomic considerations. In some countries, Black rhinoceros are excluded because of excessive illegal hunting pressure, which has seriously exposed them to highly endangered animals. This is also stated in the Convention on International Trade for Endangered Species, of which many countries are signatories.

Some vertebrates, such as the giraffe in Tanzania, the zebra in Botswana, and the eagle in the United States, are considered national games, so commercial hunting of such animals may be prohibited. Though they are part of wildlife, wild plants, and invertebrates are precluded from this concept due to the absence of intended trophies that most hunters tend to pursue.

Invertebrates' small, unimpressive trophies also exclude them from game species. However, some reptiles and birds belong to this category. For example, crocodiles and pythons have been extensively hunted for their flesh and skins as trophies; the same applies to ostriches.

The game species led to the coinage of the name Great Five Animals in the Wilderness, which refers to the animals that are difficult or frequently pursued. Unfortunately, some of the animals in the Big Five, particularly black rhinoceros, are not permitted to be hunted.

It is safe to demonstrate whether game species are only for hunting tourism or can also be used for other tourism activities. Depending on the category of the wildlife area and countries where the game species are found, the game species can equally be used for non-hunting tourist activities such as photographing, walking safaris, ballooning, and game observing.

Since it is not hunted, it is no longer a game species. In that context, descriptions of whether the animals are game or wildlife depend also on the status of the area where the animals are, sometimes the country. Let's think of the wildebeest migration between Serengeti National Park and Masai Mara National Reserve. Such migration crosses the international boundary between Tanzania and Kenya, encompassing two countries.

Similarly, migration navigates various categories of protected and unprotected landscapes. It involves various categories of wildlife areas, including Serengeti National Park, where hunting is strictly prohibited, and the migration transects open areas of Tanzania, where licensed resident and tourist hunting is lawfully permitted.

Moreover, the wildebeest migration ends in Kenya, where the conservation policy does not authorize hunting at all. It sounds odd as the same animal has different identities in different nations and different landscapes, as wildebeest becomes a game species in Tanzania but a wildlife species in Kenya, also a wildlife species in the Serengeti National Park but a game species outside the park in Tanzania.

Additionally, when game species are in areas where hunting is not lawfully permitted, they can be used for other tourism purposes, just like normal wildlife species. However, some protected areas retain both hunting and non-hunting tourism activities taking place concomitantly.

Selous Game Reserve in Tanzania demonstrates that certain areas have been isolated for photographing game species and other areas for hunting. This evidence shows that the same animal in the same protected areas with various names, the African lion, becomes a game on the pursued landscape and normal wildlife in the photographing region.

However, the word game species appears limited and biased because it was frequently used in the past, so some new conservationists can hardly describe and differentiate it precisely from wildlife. In some administrative districts and regions, the officers still retain the name, such as a colonial rank "game officer," because their primary duties as game officers were the issuance of game hunting licenses, problem animals' control, and surveillance of hunting licenses in their respective areas.

It is safe to clarify that there is nothing wrong with retaining the old name, but what is truly important is the recent review and modification of the old roles for the current game officers by broadening the tasks to also include the conservation of other non-game species, ecological monitoring, research, and education, among others.

In the context of wildlife tourism, an intelligent question may arise from someone who would like to understand whether a game species is better than a wildlife species. The convenient answer to this question would be yes, but why? First and foremost, most game species are easier to locate than wildlife species due to their enormous body sizes and activeness.

Second, their body parts, particularly the trophies, are attractive enough to entertain hunting and non-hunting tourism activities sufficiently. Their body parts, particularly the ivories of elephants and the skulls of African lions, can readily attract tourists to pursue such animals and also to observe the animals when freely ranging in the wilderness. On the other side, the issue is contradictory, as hunting species are part of wildlife species.

But the question is rephrased as "Is it possible to have game species only without other species in the protected areas?" Yes, it is inherently feasible. I have explained the context of Kitulo Plateau National Park in Tanzania and some protected areas around the world where game species are unavailable. However, artificially, it is not advisable because the existence of game species depends on other species and vice versa.

For instance, the African elephant, as a game species, is frequently cited as an ecosystem engineer, as it aids in shaping and reshaping the habitats for other animals to survive, including excavating water holes during arid seasons for other animals to consume. Similarly, African elephants frequently feed on vegetation, and their survival depends on the ecological functions of other small organisms, such as earthworms and microorganisms.

However, this situation currently occurs as some protected areas gradually lose their game species due to overhunting, climate change, human–wildlife conflict, commercial poaching, and commercial trafficking. Disrupting the ecosystem may have several implications for wildlife species, wildlife tourism, protected areas, and the socioeconomic situations of neighboring communities.

Chapter 2
Wildlife Tourism and Protected Areas

Abstract This chapter demonstrates the relationship between protected areas and wildlife tourism and protected areas. Protected areas are usually beneficial and significant as tourist destinations, attractions, and home to wildlife species. The balance between the development of tourism and the conservation of protected areas is crucial for the maintenance of the ecological integrity of the protected areas. Wildlife tourism offers outstanding ecological experiences by relying on watching wild plants and animals in their naturally occurring environment. The protected regions provide outstanding possibilities for wildlife tourists to experience and fully understand the natural ecological processes occurring in the forms of predation, distant migrations, and wilderness adaptions. Protected landscapes act as settings or platforms for wildlife tourists to directly and indirectly utilize the wildlife resources in numerous capacities, including photographing, trekking, and air ballooning. Occurring within varied designations and objectives, the protected landscapes are technically vital for conserving cultural heritages, ecological integrity, and environmental services. The categorizations and administration of protected areas might change depending on local, national, regional, and global conservation regulations. In other words, the protected area could be managed by either local communities, business enterprises, or governmental conservation authorities. The protected areas may also protect the habitats with different descriptions, including freshwater habitat, terrestrial habitat, and terrestrial and marine ecosystems. It is also possible for protected areas to conserve all three ecosystems, Saadani National Park, in Tanzania, demonstrates the circumstances. The International Union for Conservation of Nature classifies protected areas internationally to encourage the management and coordinated conservation activities globally. The international classification of protected areas is vital for avoidance of the confusion of categories among protected areas internationally. Since protected areas make up the heart of wildlife tourism, their protection is inevitable. Effective protection and administration of protected areas involve not only a coordinated effort but also trained and motivated individuals, financial stability, relevant research, and suitable institutional governance among others. Human–wildlife conflicts, climate change, commercial poaching, and unregulated tourism activities still remain the major challenges affecting the future of protected areas.

A. A. Mamboleo, *Sustainable Wildlife Tourism*,
https://doi.org/10.1007/978-3-031-97795-4_2

Keywords Wildlife tourism · Protected areas · Biodiversity conservation · Ecological connectivity · International classifications · Tourism development

Wildlife tourism is the practice of traveling for at least 24 h to observe and appreciate wild plants and animal species in natural ecosystems or protected sanctuaries. Conventionally, wildlife tourism entails traveling to natural areas or wilderness locations for tourists to appreciate the natural areas and associated species of wild animals and vegetation. It involves understanding natural animal behavior and interactions.

It is important to clarify that the most important factor is the natural behavior manifested by either plants or animals. Visitors are drawn to the behaviors of insects, animals, reptiles, and fish. Such behavior becomes natural because it occurs with a limited influence from humans on animal or plant behavior.

Wild animals and plants differ from domestic animals or plants raised in zoological or botanical gardens because the owners of the gardens usually manipulate the behaviors to satisfy the needs of their customers, and therefore their animals lose their natural behavior and become naturally unattractive.

Despite being a subdivision of wildlife tourism, the zoological and botanical gardens become less attractive than natural wildlife areas. Animals in gardens rarely express their natural behavior due to limited space, sustenance, and an unconducive climate. Some animals are underfed and diseased, which makes them unattractive to visitors to the zoo. Such manifestations are common in the developed world, where the owners of the gardens have limited budgets to feed the animals.

The African elephant illustrates the situation where the management and maintenance costs of some animals in zoological gardens may become challenging and therefore impact animal health and attractiveness. Managing an elephant in zoological gardens is costly, as it is usually active for 18 h, consuming more than 300 kg of food and drinking more than 400 l of water every day.

Having an African elephant in the urban zoological sanctuaries may become a problem. It is a challenge for many zoological enclosures, as obtaining such a quantity of vegetation and water every day to serve an elephant may be impractical, especially in arid areas and during dry seasons, when water becomes scarce and vegetation dries.

It implies that wildlife is the fundamental ingredient of wildlife tourism and the primary attraction, but it is not the only constituent of wildlife tourism. That is to say, modern wildlife tourism includes admiration of wild plants and natural habitats of wildlife and indulging in tourist activities taking place in the wildlife areas. That is to say, apart from viewing and hunting wild animals, wild plants and habitats are also part of wildlife tourism.

In other words, wildlife tourism incorporates active and passive wildlife-based activities. The typical tourism activities taking place within the wildlife areas may similarly be regarded as wildlife tourism activities because it is the wildlife that directly and indirectly influences the activities to take place at the destination.

Climbing Mount Kilimanjaro, located in Kilimanjaro National Park, exemplifies the situation of wildlife tourism activities in various aspects.

First, it is the primary home of temperate and tropical wild plants and animals, so climbers appreciate the variety and beauty of the amalgamation of tropical and temperate as they descend the mountain. Second, it is found within the natural wildlife landscape categorized as a national park, which makes the mountain a resource for mountaineering activity as another form of wildlife-based tourist activities in the park.

The primary purpose of mountaineering is to ascend the mountain and summit, but the essential argument with Mount Kilimanjaro is that the activity takes place within the natural environment of Kilimanjaro National Park, which asserts that wildlife tourism occurs in various places and forms apart from game viewing and hunting.

To simplify, wildlife areas include legally protected and human landscapes. In other places, wildlife tourism may take place in protected and unprotected wildlife areas. For a clear understanding of wildlife areas and wildlife tourism, it is necessary to have a slight illustration of protected areas. By the way, what is a protected area? How does it differ from unprotected areas, and how do protected areas relate to wildlife tourism?

Wildlife Protected Areas

The entire concept of protected areas aims at the legal exclusion of humans from illegal utilizations of resources by making wildlife areas undisturbed and uncultivated by humans. However, the effectiveness of the concepts depends on protectionists, conservationists, and anthropologists' perspectives. Since the book is not purely for conservation, the discussion on protected areas will attempt to answer the concerns raised in the previous section.

The IUCN Commission on National Parks and Protected Areas, which is technically responsible for promoting the establishment of the global network of terrestrial and marine protected areas, defines protected areas as "a clearly defined geographical space that is recognized, dedicated, and managed, through legal or other effective means, to achieve the long-term conservation of nature with associated ecosystem services and cultural values."

The above definition is technically comprehensive, as it includes a variety of essential phrases and concepts. It is important to go through each of the phrases for a clear understanding of the definition of a protected area. But before we proceed further, the definition is well-defined by the Convention on Biological Diversity.

The first phrase of the IUCN definition, "a clearly defined geography space," demonstrates the fact that for any wildlife area to acquire protected status, it must occupy a well-defined geographical space of areas of water and land, both with well-defined demarcations showing its location described in a point or polygon with associated attributes such as area management category and governance type, among others.

Protected areas must be areas of land, known as terrestrial protected areas, or areas of water, known as aquatic protected areas. To simplify, all wildlife areas occupying the land surface are referred to as terrestrial protected areas, whereas those located in water areas are referred to as aquatic protected areas.

However, the aquatic protected areas may further be subdivided into marine, freshwater, or brackish protected areas. The areas in marine ecosystems where salinity is prevalent, such as the Dar es Marine Reserves Systems, are called marine protected areas, but the ones in freshwater ecosystems, such as Rubondo and Saanane Islands National Parks, are inland freshwater protected areas.

There are those within brackish water that are neither saline nor fresh brackish protected areas. The Kilwa-Mafia Wetland Ramsar site exemplifies this type of protected landscape. With clearly defined areas, the wildlife areas must have clearly defined borders in geographically referenced areas to demonstrate their geographical territory.

Also, the phrase "dedicated and managed, through legal or other effective means," in the IUCN's definition, entails that for any wildlife area to qualify as a protected area, pertinent legal or other effective means should govern its management. Other effective measures may include a legal and institutional framework for managing protected areas.

The existence of legal and policy frameworks is essential for the effective management of protected areas. Management decisions and the use of protected areas are governed by either governmental or customary laws. Furthermore, local, regional, and international agreements such as RAMSAR conventions, CITES, and CBD exist.

In that case, relevant policy and law are essential interrelated instruments for the successful accomplishment of the "long-term conservation of nature and associated ecosystem services and cultural values." Local or international laws may further specify responsible organizations or mechanisms for the effective management of protected areas.

As a result, many protected areas are managed by special conservation agencies such as SANPARKS, which manages all national parks in South Africa; TANAPA, which manages all national parks in Tanzania; and KWS, which manages all wildlife areas in Kenya.

The last phrase contains long-term conservation objectives that can mostly be attained by having a pertinent legal and policy framework, but the long-term conservation goals of two essential aspects of natural and cultural resources. The protected areas are established for the conservation and preservation of cultural and natural values.

More importantly, the legal protection entails various restriction levels, where in some highly protected landscapes, few or no people are allowed to enter, while in low-protected landscapes, protected landscapes are integral parts of communities by allowing controlled consumptive utilization such as firewood, fruits, vegetables, and medicine collection.

Protected areas, in other words, are purposely established for the conservation, maintenance, and restoration of representative habitats in terrestrial and marine ecosystems to ensure their long-term viability, biodiversity, ecosystem productivity, and natural resources that provide socio-economic benefits in areas of special biological, ecological, educational, scientific, historic, cultural, recreational, archaeological, aesthetic, or economic value.

Most of these values are directly related to tourism because most of the resources, like archaeological, aesthetic, ecological, educational, historic, and biological values, are also termed tourism resources. Attractions are classified as the five fundamental elements of tourism.

In the above explanations, an attempt was made to answer two previous questions with some examples, but the answers may remain ambiguous if we don't elucidate the distinction between wildlife-protected areas and other protected areas. Differentiating between wildlife-protected areas and other protected areas may be simple if the actual resources in the protected areas are explicitly established.

For example, some would like to know the difference between forest-protected areas and fauna-protected areas. As per the definition of wildlife as expressly elaborated in Chap. 1, all protected areas can be wildlife areas, but if we regard wildlife as just wild animals or untamed vegetation as forest, then wildlife-protected areas and forest-protected areas are two distinct categories of protected areas.

However, such differentiation can sometimes create a problem because the forest is one of the vegetation types within the protected wild plants, as are grassland, woodland, and shrubs. Also, many wildlife-protected areas contain various forms of vegetation, including planted and natural forests. This applies to forest-protected areas; they have untamed animals that are responsible for seed dispersal and pollination. In that case, all of them can be protected areas without the specification.

We can further categorize the protected areas into publicly owned, communally owned, and privately owned protected areas, where the private ones are owned by private persons and the public ones are owned by the government. According to Palfrey et al. (2022), privately owned protected areas are as essential as public protected areas because they protect nearly 1.2% of critical global wildlife areas.

Private protected areas account for 3.4% of land under protection; therefore, they increase the global protected area network by 7.05%. It may sound peculiar to some people to hear the phrase "private protected areas," particularly in countries like the Republic of Tanzania, where most of the protected areas are publicly owned. In countries like South Africa, it is common to have privately owned protected areas.

According to the IUCN, the first concerted effort to elucidate the meaning and coverage of global protected areas took place in 1933 at the International Conference for the Protection of Fauna and Flora in London. Before that, each country had its own definition of the protected areas, inventory, and management approach.

As a result of international efforts, UNEP reports that the global network of protected areas has increased since 2010, with 22.5 million km^2 (16.64%) of the terrestrial ecosystem and 28.1 million km^2 (7.74%) of the marine ecosystem, an increase of over 21 million km^2 (42% of the current coverage).

Also, by 2021, the global database of protected and conserved areas, under the World Conservation Monitoring Centre of UNEP, recorded about 261,766 officially recognized protected areas, encompassing over 15% of the terrestrial surface and 7.4% of the marine ecosystem. Given the percentage contribution, each nation contributes variously to the global network and system of protected areas.

For example, Tanzania has allocated 38% of its surface area under protected status, which includes about 14.5 million hectares of different types of protected areas, including national parks and game reserves, which puts the country in the top 20 countries with the largest proportion of their total land area in protected areas.

According to the WorldAtlas (2023), the foremost country with a vast land surface devoted to the preservation of wildlife resources is New Macedonia (61.3%), followed by Slovenia (54.5%) and Germany (48%). Further, the World Atlas affirms that Namibia is the first country in Africa to devote 43% of its land to protection status.

The World Atlas stipulated that Namibia was also the first to incorporate environmental protection into its constitution. Namibia gazetted Dorob National Park in 2010—making Namibia the only country in the world to have a wholly protected coast. In brief, this section provided a minor insight into the definition and categories of protected areas.

Establishment of Protected Areas

Understanding the key criteria for the establishment of protected areas is important; knowing why some areas qualify to be protected while others do not is vital as far as wildlife tourism is concerned. As a wildlife tourism stakeholder, it is inevitable to understate the reasons behind labeling certain pieces of land as protected areas while leaving others with substantial wildlife resources entirely unprotected.

There have been varying understandings of the legality, logic, and concepts behind the establishment of protected areas. Apart from just understanding the logic behind the establishment of protected areas, it is also crucial to understand the procedure for establishment, which may significantly vary among countries and categories of protected areas.

The knowledge gap on criteria for the establishment of protected areas confounds not only destination managers, tour operators, and tourists, but also some conservationists. The confusion emanates from the fact that some protected areas are smaller than two football fields, others have deficient visible wild animals, and others are entirely covered by the water surface, but all of them are still referred to as protected areas.

More curiously, the surface areas of some protected areas are larger than those of certain countries. As an illustration, the surface area of Serengeti National Park is perceivably larger than Denmark. Similarly, some protected areas are 100% owned by a private or communal society. It is still unclear whether the necessary criteria for establishing protected areas exist.

We have had an extensive discussion about protected areas, which are wildlife areas under protection status where unlawful and destructive actions are severely prohibited. More importantly, protected areas are at the heart of wildlife tourism, where wildlife resources are abundant, safe, and scientifically protected. In brief, the establishment of protected areas is a prolonged procedure involving various national and international institutions before acquiring gazettement status.

The establishment procedure may be bureaucratic and lengthy for communally and privately owned protected areas, but in some countries, such procedures are friendly to privately owned protected areas as a way for the government to attract community and private sector involvement in wildlife tourism activities.

As a part of the establishment of the new or upgrading or downgrading of the protected areas, the governments or conservation agencies may undertake ecological and socio-economic studies to identify legal, cultural, economic, social, environmental, and institutional requirements to understand the existing conditions of the prospective protected area.

The topic of how protected areas are established is lengthy and intricate; let me attempt to make it simple and very related to the main theme of wildlife tourism. Apart from other considerations, protected areas are established to achieve the primary objective, which is to administer exceptional resource values within the protected area.

Among other things, the most important factors typically considered when establishing protected areas are significant features, representativeness, and feasibility. However, studies demonstrate that criteria for establishing protected areas can entirely vary depending on the reason for the protection and the owner of the protected areas. Some of the common criteria include contribution to cultural and socio-economic contributions, representativeness and comprehensiveness, natural significance, and national policies.

It is an exceptional feature or unique resource value illustrating protected areas' natural or cultural uniqueness or outstanding features, ecosystems, and landscapes at local, national, regional, continental, or global levels. The presence and extent of natural, physical, socio-economic, and cultural features in the prospective protected area determine its significant value.

Before the establishment of the protected areas, the countries usually undergo several scientific and socio-economic studies to comprehend and justify the uniqueness of the prospective protected areas in terms of natural and cultural heritage. The conservation agencies would like to prove beyond doubt the significance of establishing protected areas because the establishment and management of protected areas are economically and socially costly.

In many instances, the significance of protected areas turns out to be the primary objective for the establishment of the protected areas. Serengeti National Park's significance lies in its maximum diversity of carnivores. The significance is one of the primary conservation objectives of the park, "*the preservation of the diversity of carnivores*".

Conservation and management of a high diversity of free-ranging carnivores is one of the reasons why Serengeti National Park exists. It is a distinctive feature from a variety of geographical perspectives, from the local to the global. The statement justified the necessity for Serengeti National Park to acquire legal status to preserve the diversity of carnivores that are not found elsewhere in the world.

Because protected areas have multiple distinct and diverse cultural and natural heritages, they have opportunities to attract various wildlife visitors from different parts of the globe. The highest degree of natural and cultural significance entails the essential features that demonstrate the country's values for wildlife tourism.

It is technically required for protected areas' managers to have statements on several significance values condensed into a single sentence as a purpose statement. Yellowstone National Park has amalgamated its several significance values into the following purpose statement:

> Yellowstone National Park, the world's first national park, was set aside as a public pleasuring ground to share the geothermal wonders and preserve and protect the scenery, cultural heritage, wildlife, and geologic and ecological systems and processes in their natural condition, for the benefit and enjoyment of present and future generations.

Simply put, Yellowstone National Park's purpose statement states that it is the first national park to be established globally, aimed at safeguarding distinct geothermal and natural resources for the benefit and enjoyment of current and future generations. In other words, there is no other protected area with the mentioned features, whether locally, globally, regionally, or globally.

The establishment of protected areas is another reason for representativeness. It entails the extent to which an area can represent the cultural or natural resources of a particular ecosystem, country, region, or continent. The protected areas must demonstrate their representations of natural or cultural heritages that are inadequately included in the national system of protected areas.

In other words, elevating unprotected landscapes to protected area status necessitates a high degree of cultural or natural diversity or uniqueness occurring between transitional zones of distinct ecosystems. These are unprotected features that occur at the same site but are located across the region.

Protected areas can similarly be established for public enjoyment and offer various cultural, social, and economic benefits to local communities and governments. Protected areas offer public enjoyment in terms of leisure and tourism and provide socio-economic benefits to communities, including the provision of employment opportunities, income, and foreign exchange through tourism and other cultural and natural resources-related businesses.

Also, the availability and provision of environmental services may prompt the establishment of protected areas. The strategic location and uniqueness of the protected areas may offer enjoyment to certain areas and attract governments or private entities to establish protected areas for the enhancement of business and the garnering of foreign exchange. Some sites are strategically located.

Studies suggest that protected areas should have a large enough surface area to preserve resources and accommodate enough visitors without any environmental degradation. However, the literature hardly suggests the minimal surface area required for the establishment of protected areas. Certain protected areas, including

Pande Game Reserve and Saanane Island National Park in Tanzania, cover less than 16-km^2.

Some literature disregards the size of surface area as one of the essential criteria for the establishment of protected areas but suggests the use of ecological health as one of the most important criteria. According to the literature, it is beneficial to have small but well-interconnected protected areas rather than one large but geographically isolated protected area. The well-interconnected protected areas sustain constant wildlife movements and the genetic exchange of wildlife species.

Furthermore, protected areas may be established to protect certain unique resources, representative areas, or ecological processes. In that case, the coverage of protected resources or ecological processes, such as animal migrations and hydrological regimes, may determine the scale of the protected area.

Some exceptional features, including endemic wildlife species and natural monuments, cover tiny areas of less than five hectares. The discovery of an endemic and ovoviviparous Kihansi Spray Toad, *Nectophrynoides asperginis*, in 1996 in Tanzania required the establishment of a partial game reserve to protect an area comprising nearly 2.5 hectares of Kihansi Gorge as a partial game reserve.

Similarly, in the East African region, wildlife migration between the Serengeti National Park and Masai Mara National Reserve necessitates the establishment of transboundary protected areas to protect one of the earliest ecological processes in the world from unsustainable anthropogenic activities, *the great wildebeest migration*. Such decisions may establish one of the greatest protected areas in Africa.

Management of the protected areas should be both financially and technically feasible; the protected area is purposely established for the long-term conservation of wildlife resources, which must also be accessed and utilized by the next generations, year in and year out. In that context, the viability and durability of the protected areas become explicitly essential as far as sustainable wildlife tourism is concerned.

The endurance and viability of the protected areas depend on reliable funding and personnel in the prospective protected areas. It is difficult to operate protected areas if the conservation agency or government lacks a reliable source of funding, sufficient technologies, and qualified human resources.

Furthermore, it is difficult for the government or conservation agencies to manage protected areas if they rely solely on funds from tourism, leisure, and recreation activities. The seasonality and fluctuations of tourism and recreation activities can impact the longevity and viability of protected areas.

Categorization of Protected Areas

The abundance of protected areas on a local, regional, and global scale necessitates the categorization or classification of all protected areas into categories that enhance simple classification and descriptions. This may also help understand the types of utilization for each category. It is still uncertain to pinpoint the entities responsible for categorizing the protected areas. Are there guidelines for categorizing protected areas?

Since wildlife resources in the protected areas are owned by a particular country, any country must categorize its protected areas, though it is not mandatory. Some countries possess a few protected areas, to the extent that categorization seems inevitable. I have previously given an example of Tanzania, which has more than 900 protected areas, so in Tanzania's case, categorization may be certain.

Additionally, the regional or international categorization of protected areas requires a mutual understanding between participating countries interested in the universal categorization. Aside from local categorization, in which a country categorizes protected areas based on its own guidelines or conservation policies, an international categorization exists and follows international guidelines and criteria.

The international categorization avoids unnecessary confusion when it comes to the naming of protected areas globally. With such criteria, the national park in Uganda will be the same as in Vietnam, not like the way it is in some countries where the national park in one country is essentially a national reserve for other countries. The adoption of international guidelines makes communication between conservationists, scientists, and tourists simple because it leads to a common comprehension of categories of protected areas.

In several instances, the categorization of protected areas is based on many criteria, including the type and extent of resource utilization allowed in the protected areas, significance features, type, and current status of resources in the protected areas, among others. Let us see how the nature and status of the resources may influence the categorization of protected areas.

The protected landscape is primarily used to preserve or conserve cultural and natural resources. Some of the areas may be purposefully established to conserve resources that are at risk of disappearance, as well as highly intact or undisturbed resources. At least one of the reasons mentioned may influence either a country or the international community to categorize or classify certain protected areas into one of these categories.

For example, *Strict Nature Reserve* is an international categorization representing all protected areas with undisturbed habitats for future uses. In that case, all protected areas with such features can be categorized into this group.

Moreover, some protected areas are purposely established to protect significant features, such as endemic or extremely endangered species of wildlife or any cultural feature. Such significant features may have local, regional, or international value that deserves immediate conservation measures. Protected areas, based on their significance, may be classified as natural monuments, national parks, wetlands, or game reserves.

However, the IUCN presents clear categorizations of protected areas in the world, which most countries have adopted and classified some of their protected areas into. It is appropriate for one protected area to have more than one categorization, as international categorization does not disqualify the regional or local categorization of protected areas. It is beyond the scope of this book to describe each of the international categorizations.

After an extensive discussion of designated spaces and their categorizations, it is apparent that the differentiation between wildlife areas and protected areas is presently clear. Furthermore, most of us now understand that protected areas are established not only for wildlife conservation or wildlife tourism but also for other reasons.

It is the reason why it is essential to specify the names of protected areas into *wildlife-protected areas* instead of just "protected areas," because they are specifically established for the sustainable conservation of wildlife resources. It is necessary to understand that wildlife tourism can take place both within and outside the wildlife-protected areas because wildlife does not exist only within the protected areas.

However, several concerns may arise: Why should wildlife occupy the unprotected landscape or human settlements when there are already specially protected landscapes set aside for it to survive? It is impractical to accommodate all wildlife within the protected areas because no protected area in the world can sufficiently meet all ecological or fundamental requirements of wildlife in a single isolated landscape.

On occasion, wildlife will go off to find nutritive food, water, and reproduction, which become limited in some seasons. That is why some protected areas dig water bores in the arid seasons to supply water to wildlife, while others provide food to wildlife during the winter when all vegetation is covered by snow.

International Categorization of Protected Areas

Protected areas have been classified differently on a global scale. Certain international organizations, including IUCN and UNESCO, have been at the forefront of categorizing protected areas. Their categorizations vary from purely scientific to just recognition of protected areas after participation or attaining certain achievements. Biosphere reserves, world heritage sites, Ramsar sites, and IUCN categories of protected areas exemplify such categorizations, involving a procedural categorization of protected areas.

In many cases, the interested protected area or country must be observed to be eligible for categorization. However, this necessitates more than just the country's consent but also an endowment of relevant features in the protected areas. The special committee is involved in endorsing the classification of protected areas.

The country or protected areas can be classified into more than one international category, as is the case for Serengeti National Park, which falls under the World Heritage Site, IUCN, and UNESCO biosphere reserve categories. The international categorizations are purposefully established for either certification or classification of protected areas.

Heritage International certification schemes, such as World Heritage Sites and UNESCO Man and Biosphere Reserves, are certified or recognized based on the uniqueness of their resources and features or compliance with certain requirements set by awarding organizations.

International categorization entails classifying protected areas based on their similarities and differences in terms of accessible resources and features. The processes involve all the protected areas around the globe. The IUCN has global protected areas in seven main categories based on resource condition, availability, and uniqueness.

The categories include strict nature reserves, wilderness areas, national parks, natural monuments or features, habitat or species management areas, protected landscapes or seascapes, and protected areas with sustainable use of natural resources.

According to the IUCN, the rigorous nature reserve includes protected areas that protect biodiversity and geomorphological features. Moreover, such areas rigorously control and limit human visitations, use, and impacts as a means of heightening the protection and conservation values of resources. The strict nature reserves function as global reference areas for scientific research and environmental monitoring.

Wilderness areas are vast, unmodified, or slightly modified protected areas that have been able to retain their natural features for a long time without permanent or significant human residents and are protected and managed for the preservation of the natural conditions. According to the IUCN, all areas with vast surface areas without human habitation are classified as wilderness. This category includes Tanzania's Serengeti National Park.

A large natural area, a national park, is set aside to protect and sustain large-scale ecological processes, ecosystem services, and complementary species in the area. It also provides opportunities for humans to experience the spiritual, scientific, educational, and recreational values of the area.

The natural monument protects specific natural monuments, such as landforms, seas, lakes, rivers, mounts, geological features, caverns, and any living features. The natural monument is very tiny in extent compared to other categories. The Victoria Falls are an example of a natural monument.

Habitat or species management area is an IUCN category aimed at protecting specific species or habitats, which can be endemic, endangered, or uncommon. Such protected areas may require regular conservation intervention to meet the needs of certain species or maintain the habitat's adequate conditions.

Protected areas with sustainable use of natural resources entail the kind of protected area that the long-term coexistence and interaction between humans and nature have resulted in. As a result, the area has excellent ecological, cultural, biological, and scenic features. That is to say, maintaining and safeguarding such interactions becomes vital for the integrity of the protected area.

Protected areas with sustainable use of natural resources are the category that aims to conserve ecosystems with cultural values in mind while also using traditional resource management systems. Such areas are huge and still in natural condition. The areas encourage moderate-level use of natural resources.

Local Categorizations of Protected Areas

In the introduction chapters, local categorizations of wildlife areas have been frequently mentioned. It is the destination's policy that usually determines the number and variety of categories of wildlife areas. With that in mind, it is difficult to enumerate and define the local categorizations of protected areas from all destinations around the globe.

Also, it is unjust to speak about wildlife tourism without the understanding of local and international categories of protected areas. The international categories have been well defined in the previous sections; this section specifically elaborates on the local classifications of protected areas by citing examples from a few countries.

The enormous benefaction of natural resources in some countries led the author to opt for specific countries as guidelines for discussion on the local categorization of protected areas. In other words, Tanzania, South Africa, and Botswana exemplify Indigenous categorizations in an African context. The mentioned countries are critical as various categories of wildlife utilization, including consumptive animal utilization, are allowed, which are suitable for the provision of such examples in the next chapters.

Tanzania has more than 900 protected areas that are locally classified into two primary categories: marine areas and terrestrial protected areas. The marine protected areas include all protected areas that are located in the marine habitat, which are marine parks and marine reserves only.

In Tanzania, terrestrial protected areas encompass all protected areas that are located in inland aquatic bodies and terrestrial habitats. These protected areas include national parks, Ngorongoro conservation areas, game reserves, marine reserves, marine parks, game-controlled areas, wildlife management areas, forest reserves, nature reserves, community-based forest reserves, open areas, and forest plantations.

The local categorizations are based on the form of utilization, management authority, and management policy. In Tanzania, the management or conservation authority has specific policies governing resource management and utilization in each category. The procedure for establishing each category may differ significantly between destinations.

In Tanzania, the National Park Ordinance No. 412 of 1959 governs the management and utilization of all national parks. Initially, the ordinance established the administration of Tanzania National Parks (TANAPA), followed by Tanganyika National Parks, which is the management authority intended to oversee more than 22 national parks in Tanzania.

The national park differs from other local protected areas in that human activities, including cultivation, livestock keeping, and settlements, are rigorously prohibited. Also, economic development activities such as mining, commercial fishing, harvesting of timber and non-timber forest products, hydrocarbon exploration, and hunting of any description are not allowed in the national parks.

The parks are exclusively for the complete protection of biological diversity and representative areas, and it is the highest category of protected areas, as most human activities are not permitted, such as human habitation and livestock keeping in the park, which is a criminal offense. However, national parks allow non-consumptive recreational, tourism, educational, and research activities. Furthermore, local people may be allowed to enter and conduct their cultural activities, including worship.

Commercial or domestic fisheries, deforestation, quarrying, and many other consumptive activities are not permitted in the parks. It is essential to emphasize that each national park was purposely established to safeguard biological, natural, cultural, and socio-economically significant or distinct resources.

The presence of significant resources that attract tourism activities makes each park individually unique. For example, the high diversity and abundance of carnivores make Serengeti National Park locally, regionally, and globally unique and exceptional. Therefore, such richness attracts both national and international concerted efforts to rescue the resources from avoidable disappearances.

The uniqueness of such resources draws many tourists from all over the world to the park. According to this concept, a national park is established to preserve identified exceptional resource values. The objective of establishing a national park is to ensure that all resources in the park, including the most significant ones, are resolutely protected and their consumptions comply with the stipulated TANAPA's policy.

Game reserves are another local category of protected areas in Tanzania. The nomenclature of game reserves was inherited from the colonial regime for reasons provided in the first chapter of this book, namely that the colonial administrations allocated such a name to protected areas designated for the conservation of enormous concentrations of game species.

Even some of the existing and famous protected areas used to be game reserves before being elevated to their current protected status. For example, Serengeti National Park, Ngorongoro Conservation Area, and Ruaha National Park initially used the game reserves.

In Tanzania, apart from trophy hunting, game reserves are also established for the conservation and management of significant features. In those areas, another consumptive wildlife utilization is rigorously prohibited. Livestock keeping, cultivation, quarrying, tree toppling, and human habitation are not permitted in the reserve.

Such restrictions are essential for the conservation of biological, natural, and cultural resources. Like many other protected areas, licensed research, education, and scientific activities are allowed. Some game reserves, such as Mpanga Kipengele and Selous Game Reserves, allow wildlife photographic activities.

The Tanzania Wildlife Administration Authority is the conservation agency responsible for managing all wildlife reserves in Tanzania. It is the same agency that administers all wildlife outside the national parks, Ngorongoro Conservation Area, and privately owned and communally held protected areas. Also, the wildlife reserves are regulated by the Wildlife Conservation Act Number 5 of 2009.

TAWA administers more than 25 game reserves in Tanzania, including the famous Selous Game Reserve and Rungwa Game Reserves. The Ngorongoro

Conservation Area is a local classification of protected areas designed for multiple uses within the territory. The three essential pillars of Ngorongoro—anthropogenic use, conservation, and tourism—make the protected area locally, regionally, and globally unique. It is a secure location where coexistence between humans and fauna occurs.

In the Ngorongoro Conservation Area, human settlement and livestock keeping are authorized only in certain parts of the protected areas. Non-consumptive wildlife tourism activities, including game observation, walking safaris, and trekking, are allowed. Educational and research activities are also permitted.

It is essential to clarify that livestock keeping and human residents are permitted only for a few designated ethnic groups, not for everyone. Such ethnic groupings are believed to have less impact on fauna. However, overpopulation from ethnic groups appears to cause ecological problems, so the government of the United Republic of Tanzania has asked some of its residents to voluntarily abandon the protected area to save biodiversity.

Resident hunting and trophy hunting are not permitted in the Ngorongoro Conservation Area. Mining, commercial and subsistence farming, as well as harvesting timber and non-timber forest products, are all strictly prohibited in the protected area. The Ngorongoro Conservation Area Authority is the conservation agency responsible for managing and conserving the Ngorongoro Conservation Area.

Part II
Sustainable Wildlife Utilizations

Chapter 3
Wildlife Tourism Utilizations

Abstract This chapter discusses several kinds of licensed wildlife utilisations. Wildlife tourism involves consumptive and non-consumptive utilisations of wildlife resources. It is important to highlight that both consumptive and non-consumptive utilization may contribute greatly to tourists' enjoyment but also to economic, cultural, and environmental conservation when executed responsibly. It is possible to maximize the advantages of consumptive and non-consumptive wildlife utilizations while adhering to established social, economic, cultural, and environmental standards. Wildlife tourism depends on protected landscapes, which is a core region offering in situ conservation for wildlife species while excluding and constraining how humans may interact or access the natural environments. The networks of protected areas vary among countries, where each country has a variety of kinds and sizes of protected areas offering different types of wildlife utilisations. Countries such as China, India, Brazil, and Tanzania demonstrate the usefulness and significance of protected areas to the development of wildlife tourism. The Tiger Reserve in India contributes substantially to the massive tourism earnings. The Yunnan province of China, which offers a range of beautiful wild plants and animals, including the Asian leopard, implies the popularity of wildlife-based travel. The Amazon rainforest, a worldwide significant biodiversity hotspot with an insufficiently developed tourism potential, is one of the world's attractive wildlife tourism attractions. On the other side, Costa Rica highlights the relevance of protected areas in ecotourism development by designating up to 27% of its territory for wildlife tourism. Similarly, Australia and the United States spend enormous surface areas of their land on the preservation and management of animal resources, which attract domestic and international wildlife tourists. Some African countries continue to maintain their popularity as important wildlife tourism destinations by supporting the protection of appealing and endemic wildlife species. The Serengeti-Masai Mara environment has sustained the greatest and arguably the oldest nomadic migration of wildebeest. The protected areas provide opportunities for wildlife tourists to indulge in non-consumptive utilisations such as game viewing, kayaking, and photographing, among others. On the other hand, some protected areas provide tourists with the opportunity to engage in consumptive utilisations, which include tourist hunting, therapeutic hunting, domestic hunting, and traditional hunting. However, both mis-

A. A. Mamboleo, *Sustainable Wildlife Tourism*,
https://doi.org/10.1007/978-3-031-97795-4_3

managed consumptive and non-consumptive utilization can be harmful if unregulated. Addressing concerns like human–wildlife conflicts, climate change, commercial poaching, and sustainability in wildlife utilizations is crucial for the long-term profitability of wildlife tourism.

Keywords Wildlife tourism · Consumptive and non-consumptive utilization · Ecological sustainability · Biodiversity conservation · Community involvement · Adaptive management strategies · Responsible tourism

Wildlife tourism is purposeful travel to wildlife areas to enjoy the natural areas and animals. The previous chapters have intensely explained wildlife tourism by defining wildlife species, game species, and wildlife-protected areas, which are important parts of it. However, it is important to understand that the definition of wildlife tourism depends on various contexts, including national and international wildlife policies, ecological underpinnings, socioeconomic contexts, and perceptions of wildlife tourism stakeholders.

In 2002, Orams concluded that apart from offering psychological, social, and economic benefits to humans, wildlife feeding activity in natural wildlife habitats adversely affects the wildlife psychologically, socially, and economically. Therefore, this is one of the factors determining the description of wildlife tourism and its activities. UNWTO defines wildlife tourism as the direct and indirect observation and interaction with native animal and plant life in their naturally occurring habitats.

This universal definition put in place by UNWTO implies that wildlife tourism involves total interactions with undomesticated organisms, not the alien invasive species or domesticated animals in our homes, zoological, and botanical gardens. The majority of literature and scholars define wildlife tourism as similar to the UNWTO. However, the definition may sound controversial and unfair in certain contexts.

Assume the tourist traveled for thousands of miles to specifically observe the domesticated or caged species off the protected landscape, which is both endemic and extinct in the wilderness, meaning that it is only found at a specific destination. Such travelers fit into all the characteristics that make a tourist; however, it is interestingly unclear in which category of tourism the tourist might be placed. According to the definition, such tourists are not considered wildlife tourists. However, this would not be the case for certain destinations.

The zoological and botanical gardens have been the reservoirs of many endemic, extinct, and endangered wildlife species. Apart from educational and recreational activities, such facilities contribute directly and indirectly to in situ and ex situ wildlife conservation of wildlife species. In countries with enormous and diverse wildlife resources, zoos and botanical gardens usually remain as educational, entertainment, and scientific facilities.

Nevertheless, in certain countries such as Australia and New Zealand, zoological captive wildlife is a serious attraction for wildlife tourists. People travel for more than 24 h to observe and interact with the animals in zoos. Australia had more than

100 zoological gardens in 2001, attracting more than eight million international and domestic wildlife tourists.

International wildlife tourists are one-third of visitors to the zoological gardens, contributing substantially to the economy of the country and the creation of employment opportunities. It is technically unjustifiable and unrealistic to underestimate or exclude the contribution of domesticated species of wild animals to wildlife tourism by disregarding their activities from wildlife tourism.

Despite the uncertainty of whether wildlife captives or zoos are part of wildlife tourism, it traditionally takes place on public, communal, and private lands. The idea of establishing protected areas is ancient and universal. Various continents established protected areas for special interests.

The literature points out that Europeans started the protected areas only for the exclusion of the poor from the use of royal hunting grounds set aside for the rich and powerful people about a million years ago. In certain parts of Africa, the protected areas were set aside for protecting cultural and natural resources with special uses as sacred groves.

With time, these special sites allowed the public to access them, providing opportunities for wildlife tourism development globally. The protected areas serve as the backbone of wildlife tourism, as most of the activities take place in the communal, public, and private protected landscapes. WTTC reported that China has about 120 protected areas covering 5% of marine and 16% of terrestrial areas.

China has locally categorized the protected areas into National Forest Parks, National Natural Reserves, National Key Scenic Resorts, National Natural Reserves, National Geoparks, National Wetland Parks, National Mining Parks, and National Water Reserves.

These are the natural areas of land and water dedicated by the government of China for wildlife tourism to take place. Currently, China has many internationally recognized protected areas; it is one of the countries with at least 1865 protected areas classified as national parks by the IUCN.

However, the Yunnan Province seems to dominate the potentiality of wildlife tourism as it hosts the greatest wildlife diversity and population in China. According to the WTTC, the province hosts 55% of all of China's vertebrates and about 200 endangered as well as rare species. Some of the notable species in the province are the Asian leopard, wild ox, hornbills, Yunnan golden monkey, and Asian elephant.

In the past, the performance of the wildlife tourism industry in China was relatively unsatisfactory due to unsupportive rural tourism infrastructure and superstructure. The undeveloped rural infrastructure where most wildlife tourism activities and operations take place limited the accessibility of wildlife attractions by wildlife tourists.

Despite being one of the countries with many internationally recognized protected areas, access to Chinese wildlife resources was not easy or convenient because of visa restrictions for international visitors and insufficient human and organizational resources in the areas of wildlife tourism and hospitality.

However, the flexibility of international Chinese policy, massive infrastructural and superstructural rural investments, and modernized education systems have attracted international wildlife tourists to conveniently access the escalating number of wildlife attractions in China. It is important to acknowledge that wildlife tourism in China occurs as ecotourism, connecting culture and the natural environment.

According to the WTTC, the Chinese perceive ecotourism as the connection between nature and humanity, aiming at connecting humans' present and past attributed to the concepts of Confucianism, Buddhism, and Taoism. The government of China promotes the alignment of wildlife tourism with sustainable tourism development by complying with cultural and socio-economic principles. The government intends to measure the future success of the protected areas based on the implementation of sustainability principles.

Also, India is one of the world's biodiversity hotspots, ranking 12th globally for hosting diverse and enormous wildlife. Such diversity makes India among the most famous wildlife tourist attractions globally and perhaps the best wildlife tourism destination in Asia. Studies show that the majority of wildlife tourists travel to India primarily to see and interact with tigers.

WTTC reported tourism as one of the largest foreign exchange earners of India, contributing more than 3.6% to the GDP. The country is home to 7% of mammals, 62% of reptiles, and 6% of wild plant species in the world. The protected areas include 672 protected areas covering about 6% of India's terrestrial surface area.

Brazil is a strategic wildlife tourism destination because of the exceptional endowment of wildlife resources it possesses. It is the continental representative of wildlife resources, as all wildlife species occurring in all countries of South America are found in Brazil.

Interestingly, 60% of the Amazon forest, including the important breeding areas for the jaguar, deer, otters, tapirs, and giant anteaters, lies in Brazil. It has the highest biodiversity in the world; at least two-thirds of all species in the world are found in Brazil.

The country has 2309 protected areas covering almost 19% of the surface areas and covering approximately 27% of its marine aquatic ecosystem. Brazil has 334 federally managed protected areas spanning around 170 million hectares. Before the COVID-19 pandemic, Brazil recorded about 15.3 million visits to its attractions.

Certain protected areas have attained international recognition because of their involvement in environmental and sociocultural commitments. In those aspects, 27 protected areas in Brazil are internationally recognized as Ramsar Sites, 7 protected areas are World Heritage Sites, 7 areas are Biosphere Reserves, and only one protected area is internationally categorized in the Geoparks category. Interestingly, studies affirm that each dollar invested in a Brazilian protected area returns seven dollars in economic benefits.

Brazil is internationally recognized because of its exceptional endemism; it is only second to Indonesia. Despite the endemism that brands Brazil as one of the most key areas for wildlife tourism on the planet, the contribution of the Amazon ecosystem to Brazil remains insufficient as it contributes only 1% to regional GDP.

Wildlife tourism in Brazil occurs in terms of wildlife observation and interaction with nature. In most cases, wildlife tourism in Brazil takes place in state-protected areas; the country receives more international than domestic wildlife tourists practicing ecotourism, adventure, and wildlife viewing are the main activities of wildlife tourists in Brazil. However, the COVID-19 pandemic stimulated the domestic tourism market in Brazil.

While Costa Rica hosted about five million people in 2017, 2.1 million domestic and international tourists visited, half of them being domestic tourists. It is one of the countries with the highest domestic tourist visitations. With about 187 protected areas covering over 27% of the country's surface area, Costa Rica hosts 5% of global biodiversity, making the country one of the most diverse in the world and the most popular ecotourism destination on the planet Earth.

Wildlife tourism in Costa Rica contributes 12.9% to the country's GDP and creates one in every eight jobs. Peaceful and smooth alternation of political power makes Costa Rica one of the safest wildlife tourism destinations. Also, integrated and well-developed tourism superstructure and infrastructure brand Costa Rica as an easy-to-reach and easy-to-travel wildlife destination. More than 40% of visitors to Costa Rica are from the United States; this is enabled by the geographical advantage of Costa Rica being closer to the United States.

The United Kingdom National Parks categorize the protected areas into Areas of Outstanding Natural Beauty and National Scenic Areas. The Kingdom has 49 Areas of Outstanding Natural Beauty, covering 19,596 km^2 in England. 844 km^2 in Wales 2861 km^2 in Northern Ireland. The Kingdom has 15 national parks, covering 12,126 km^2 in England. 4141 km^2 in Wales 5665 km^2 in Scotland.

The National Scenic Areas are 40, and all are in Scotland, covering about 13,783 km^2. In the United Kingdom, birdwatching, particularly with eagles and puffins, is one of the wildlife tourism activities. Also, observation of animals like deer, red squirrels, badgers, and mountain hares takes place in the United Kingdom. Scotland is one of the famous wildlife tourism destinations in the United Kingdom; past estimates show that wildlife tourism accounted for 1.12 million annual trips, earning about £276 million in 2009.

The protected areas of the United States are managed at the federal, state, and local levels to maximize the protection efficiency, sustainability practices, and stakeholders' involvement. In 2022, the United States had 42,826 protected areas spanning around 1,236,486 km^2, equivalent to 13% of the total surface area of the country.

Moreover, the country has more than 871 marine protected areas. The surface areas covered by the US protected areas are equivalent to 10% of all protected landscapes on the planet Earth. In other words, the United States contributes 10% of its surface area to the protected area of the world.

It is important to know that the United States protected areas are managed in various administrative capacities, including local, tribal, state, and federal. Each authority has a unique way of controlling the protected area. The United States has more than 6770 federal-level protected land areas managed by various protected state agencies.

Such areas include the National Park Service and the United States Fish and Wildlife Service. The federally protected areas are locally categorized into the National Park System, National Forest, and National Conservation Lands, among others. Interestingly, such protected areas contribute to 12% of the global IUCN's Strict Nature Reserve and Wilderness.

The state-level protected areas occur in various sizes, while others allow human settlements in the park. For example, several towns are located inside the Adirondack Park. The local-level protected areas range between nature reserves and picnic sites. Various authorities own and manage the protected areas, including cities, districts, and metropolitan authorities. One of the largest locally owned protected areas in the United States is South Mountain Park, which is in Arizona.

Such protected areas' diversity and massiveness reflect the traditional outdoor enjoyment of the United States. Both local and international wildlife tourists have opportunities to access and enjoy the natural wildlife resources spread over the large area of the country. The large population and stable influx of foreigners make the United States an essential wildlife tourism destination.

It is estimated that US National Parks recorded 330.9 million visits in 2017. Domestic wildlife tourism made about 258 million trips. The majority (75%) of domestic wildlife tourists visited public and private lands while spending at least US$11.6 billion in 2016.

In Africa, wildlife is the single biggest driver for tourism growth. The United Nations World Tourism Organisation found that 80% of annual sales of trips to Africa are for wildlife watching. Wildlife is the single biggest driver for Africa's tourism growth. The United Nations World Tourism Organisation found that 80% of annual sales of trips to Africa are for wildlife watching.

Furthermore, the Protected Planet found that protected areas cover 1,692,884 km^2 of Australia's land surface area. It is about 22.03% of the Australian territory. The local or national categorization of protected areas' networks in Australia includes national parks, nature reserves, marine reserves, Indigenous protected areas, and privately protected areas. Two-thirds of the protected areas in Australia belong to the strict nature reserve, which is the highest categorization of protected areas as described by the International Union for Conservation of Nature.

The government of Australia owns and manages nearly half of the protected areas. About half of protected areas fall under Indigenous Protected Areas (IPAs), which occupy at least 74 million hectares.

It is one of the countries with the largest number of privately owned protected areas, amounting to at least 6000 privately protected areas. The presence of endemic species and diverse wildlife species makes Australia a perfect destination for domestic and international wildlife tourists. The visitors are motivated by observation and interaction with terrestrial and aquatic wildlife species in protected areas, zoological gardens, and aquariums.

Nonconsumptive Wildlife Tourism Utilizations

It refers to the indirect utilization of wildlife resources from their natural environment without killing, capturing, or consuming wildlife resources from their natural areas. Depending on the type of destination or attraction, nonconsumptive wildlife tourism involves, among other things, walking safaris, mountain climbing, animal feeding, wildlife viewing, photography, film tourism, snorkeling, and scuba diving.

Furthermore, non-consumptive wildlife tourism may be separated into two different groups: directly and indirectly associated with wildlife tourist activities. Some activities are not specifically tied to wildlife tourism but nevertheless take place in wildlife regions. For example, kayaking, horseback riding, treasure-seeking, marathons, cycling, picnics, and honeymooning, among others, are indirectly tied to biodiversity.

Nonconsumptive wildlife tourism activities are licensed tourism operations performed by trained professional tour operators or the protected areas themselves. It is common for privately owned protected areas to manage all nonconsumptive activities themselves as an aspect of income generation. In some publicly owned protected areas, third-party organizations are commissioned to execute the activities.

The activities may take place in public and private protected areas, unprotected and protected landscapes, and natural and exotic tourist attractions. Due to the nature of nonconsumptive activities, many wildlife tourism operators consider it a sustainable form of wildlife utilization because it does not involve the actual offtake of wildlife resources.

The norms and laws exist for wildlife visitors and tour operators to engage in indirect wildlife tourism operations. Before the formation and licensing of any nonconsumptive wildlife tourism activity, the competent authorities may perform ecological and socio-economic studies to assess the viability of each activity.

As far as wildlife tourism sustainability is concerned, each of the planned wildlife tourist activities should be subjected to an Environmental Impact Assessment. It is a technique for detecting any possible repercussions, offering mitigation measures, and assessing wildlife visitor activities to ensure environmental integrity and the welfare of the nearby communities in protected areas.

Among others, nonconsumptive wildlife operations include wild camping, snorkeling, health tourism, hot air ballooning, bush feasting or lunching, scuba diving, wildlife filming, wildlife feeding, canopy walking, kayaking or canoeing tourism, and treasure hunting.

Wild Camping

It is a strictly outdoor recreational and adventure activity, often called bush camping, requiring wildlife tourists to build a temporary protective shelter or tents in the specified location inside the protected area or wilderness to remain overnight. Bush camping can be complicated, but if done right, it can be fascinating. Some protected areas have established specific bush camping sites, also called public camping sites, where basic services such as water supply and washrooms are available. In certain regions with deadly animals, security services may be provided.

Wild camping is a recreational wildlife activity that involves sleeping in a shelter tent or special camping vehicle overnight as a method of spending time with visitors in wildlife-protected areas to pursue their fun activities away from inhabited areas. Wildlife camping frequently occurs in groups, when friends or families come together to spend their nights fleeing the populous places and engaging intimately with the environment.

Wild camping offers wildlife tourists a temporary home by keeping the campers away from their regular homes. Sleeping in tents without power or making meals without modern cooking tools and cookers might be an innovative experience for many wildlife travelers. More significantly, resting at night while hearing the sounds of various nocturnal animals without urban automobile hooting and noises may deliver another wonderful memory to wildlife travelers. However, wild camping may incorporate various activities throughout the day, such as animal watching, hiking, and walking safaris.

Additionally, bush camping demands previous planning, although this may rely on numerous cases. In the case of a private tour, this is more important to domestic visitors on a private tour. The campers should arrange their sufficient amount of food, clean and safe water, tents, transport, first aid kits, admission charges, camping fees, and specific touring equipment, among other things, for camping inside the protected regions. Tourists go under the cover of tour operators; it is the tour operator who may arrange all the essential camping accessories on behalf of the tourists.

Though bush camping is an adventure, wildlife visitors may need to have sufficient camping abilities when camping in protected regions that lack particular approved camping grounds. Such environments require guests to properly select the camping sites by themselves, so particular skills and knowledge are essential to building shelters away from conventional paths and territories of hazardous animals. Constructing tents in such settings is risky for campers and animals.

Moreover, wild camping also comes with certain responsibilities, as conservation authorities regularly supervise the actions of campers as wildlife visitors to conserve the animals and tourists. In other words, the preventive precautions should be thoroughly studied before entering the protected areas or camping at the campground. It is necessary for campers to respect the animals, cease feeding wildlife, prevent burning, avoid entering with pets or live animals, avoid casual trash, and not camp in the forbidden areas.

Having skilled wilderness survival skills is vital while camping since anything may happen at the campground. The technique is critical for escaping any uncertainties that commonly arise in jungle camping. At least one of the campers must be practically conversant with the required survival abilities for saving the rest of the group in tough periods. Events such as snake and insect bites, running out of drinking water and food, getting lost in the huge protected regions, and wildfires are prevalent when camping and thus demand proper judgments and actions.

That reminds the campers to also have appropriate first aid skills to take care of basic medical treatments during the camp. As many of the campers may fall into health issues at any moment, with a well-prepared medical kit, the skilled first aider may assist in taking care of and preserving the lives of necessary wildlife tourists before reaching out for expert medical remedies.

The wilderness is abundant with wild plants and creatures that could cause health issues for wildlife visitors in a variety of ways, from allergies to infectious health concerns. That is, when camping, having a qualified and competent first aider is crucial.

Apart from being fun, bush camping can also be an educational activity by teaching tourists or young people critical survival or life skills. Wild camping incorporates living styles that are rarely encountered in urban settings, which may help wildlife travelers understand what is happening on the other side of life. During camping, it is truly shocking to use kerosene lights but not powered ones, to use pit toilets, to eat with hands, to cook with charcoal or firewood, and to collect water. All these may serve as valuable lessons for some wildlife visitors.

Snorkeling

It is a hydro- or aquatic adventure swimming sport, including swimming in a river, ocean, or lake for enjoyment objectives taking place in wildlife habitat. This wildlife tourist activity received its name from the breathing device that every swimmer must use while swimming, called a snorkel.

When swimming in the sea, this innovative diving mask helps swimmers breathe. Moreover, the snorkel helps the swimmer have a better view of the aquatic environment. Apart from the snorkel, the swimmer is also equipped with swim fins, which enable smooth, controlled, and fish-like swimming in aquatic environments.

Though it is a challenging exercise, it scarcely demands any advanced preparation, training, money, or energy to complete. However, protective precautions and carefulness are required while practicing snorkeling. Let's explore the link between snorkeling and wildlife tourism. In the introductory part of defining wildlife, we declared that the natural habitat of animals is arboreal, terrestrial, and aquatic.

In that context, snorkeling takes place in the aquatic environment where aquatic animals live. The ocean, rivers, and lakes are some of the aquatic habitats with diverse sorts of animal species. Observation of water animals is not as simple as that of terrestrial ones because it hardly requires wildlife tourists to have unique and specialized equipment and training, but simply eyes and feet. In an aquatic habitat, wildlife visitors need particular equipment to enjoy the unique aquatic species resources.

Depending on the region where it takes place, snorkeling may be free on public beaches, but wildlife visitors have to pay an access fee in marine protected zones. However, for an amazing animal-watching experience, snorkeling in the protected marine zones might be more intriguing than in public. The biological resources in the marine protected areas are unique and highly conserved, such as coral reefs, seagrass, colorful fish, and meadows.

Snorkeling may occasionally require water transport, notably a boat, to reach remote snorkeling sites. However, snorkeling does not take place in the whole protected area, lake, river, or ocean but in the areas where physical, biological, and chemical circumstances favor wildlife tourism. The biological state comprises the biological variety of appealing biological resources for wildlife visitors to enjoy. The biological resources in the protected regions are unevenly distributed.

Talking about the physical condition, the snorkeling place should have supportive physical qualities, including water temperature, clean water, bearable waves, and current. Since snorkeling affects human life, the snorkeling area must have warm and clean water for wildlife visitors to be able to watch the aquatic attractions. Also, low or no water waves and currents may improve wildlife visitors' swimming experience.

The chemical composition of the snorkeling area is also significant, as it impacts the type of animal resources and the swimming experience of wildlife visitors. The snorkeling site with a high or low pH may be harmful; furthermore, the snorkeling location with heightened chemical contents, for instance, the place where eutrophication is the major characteristic, can be an issue for the swimmer.

Bush Lunch and Dinner

Wildlife tourism takes place in various ways, since individual tour and hospitality specialists may see the unique value of wildlife resources differently. The hospitality specialists came up with one of the most intriguing approaches to wildlife resource utilization: bush lunch or bush supper. It has perhaps become somewhat famous in recent years as many tour companies or hotel facilities provide bush lunch, supper, or both for wildlife visitors in the wilderness.

It may be a component of a tour package or the only package that requires the visitor to dine in the wildlife areas. In other words, wildlife tourists take their lunches and dinners in the center of the wilderness, surrounded by wild flora and animals. More interestingly, the eating takes place in an appealing natural setting in the presence of fresh air.

Based on the budget, kind of supplier, and other parameters, the style of meals may range from basic or picnic-style meals to sophisticated gourmet delicacies. Such a meal surpasses the most dining experiences many wildlife travelers have ever experienced. Of course, this form of wildlife tourism is still perplexing due to the ambiguous descriptions.

Dining amid elephants, lions, buffalo, or other deadly beasts? Is it possible to even enjoy your meal? Is it legal? Where do you get the food from? Who cooks the food? In summary, it is possible to eat anywhere in the bush unless visitors observe the management planning and zoning of the particular protected area.

According to the planning and depending on the kind of tourist activity, a particular area, resource, and time have been chosen for every wildlife tourism activity taking place inside the protected area to improve the experience while safeguarding wildlife resources.

That is to say, bush lunching or eating can only take place in specified protected areas and zones. Therefore, eating at the heart of risk may depend on the management plan's specifications, since the purpose of wildlife tourism is not to promote or assist human–wildlife conflicts but to provide visitors with pleasure and respect for wildlife.

In most instances, the protected areas select and define safe yet beautiful and environmentally compatible areas for bush eating or lunching to avoid dangerous animal encounters and damage to the environment. In addition, wildlife rangers or game scouts may be employed to protect wildlife tourists from deadly creatures and ensure that the eating experience takes place as smoothly as it was intended. With the above description, it is feasible for wildlife visitors to enjoy their meals if the restrictions of the individual protected regions are skilfully observed.

According to the laws and regulations of particular conservation agencies, bush lunch or feast may be permitted or forbidden in some protected areas, as it was earlier stated. Tour operators and hotels are frequently responsible for preparing the bush lunch and supper package for wildlife visitors. The tables and chairs are used to set up the bush lunch or supper in the selected place inside the gorgeous wilderness for wildlife visitors to enjoy.

Furthermore, the type of cooked meal may vary between traditional, a blend of traditional and continental, or simply continental food. Also, food service providers may complement bush dinners with other entertainment, such as performances of traditional dances, as a method of immersing wildlife visitors in genuinely authentic wildlife environments. Such eating and performance events may include a group of friends, family, or workmates yearning for a fresh nature encounter and may sometimes be pricey in terms of money.

However, eating lunch in the wilderness may come with environmental and life consequences. If imprecisely done, any trash left unattended after meals is harmful to wild animals. As artificial food for scavengers, it may poison wild animals to death or modify the normal habits of wild animals for beggars.

Depending on the region, bush dining, and lunching may maximize human and wildlife interactions, therefore the conflicts between wildlife visitors and hazardous animals. This may result in human and animal fatalities or injuries. Bush dining incorporating fires in the dry season and on a site with an enormous fuel load may be a good cause of wildfires in the protected regions.

It is vital to note that the meals for bush dining and lunching are not produced at the dining site, but occasionally the operators may use a fireplace at night to include the sensation of wildness. Once escaped, such fire and fireplace turn into a wildfire. It is technically impossible to manage wildfires in such locations once they begin. Most of us are aware of the implications of wildfires on animal species and the natural environment.

Health Tourism

When defining the phrase wildlife in the introduction section, clarifications on the components of wildlife were also cleared. With those components, it is practicable to make health tourism a subset of wildlife tourism. Living, non-living, wild flora, and wild animals are major components of medical tourism in protected and unprotected landscapes. Health tourism is one of the very ancient components of tourism, although it has gained popularity and become widely recognized, and established lately. Medical tourism is a subcategory of health tourism, where medical treatment is the major activity.

Like other kinds of tourism, health tourism entails mobility, visitation, and spending at health tourism and other tourist establishments. It is not a new notion but a recently growing practice of tourism internationally, as, from ancient times, people used to go from their homes to different regions seeking excellent health care. Globally, at least 15 million tourists go annually overseas to receive economical, professional, and dependable medical services, mainly for dentistry, transplants, cancer treatments, and cosmetic surgery.

However, the objective of this section is not to explore the science and ideas underlying health tourism but rather to justify how wildlife resources might contribute to health or medical tourism activities in a protected environment. As a diversification strategy for wildlife tourism development, health tourism is the next category of wildlife tourism owing to the integration of animal resources for health reasons. Consciously and unintentionally, wildlife visitors may get emotional and bodily relief from animal resources.

Wildlife Resources, particularly waterfalls, are expressly recognized for increasing psychological relief for anxious wildlife visitors. Studies, like the one done by Jo et al. (2023), indicated that a waterfall boosts sympathetic nerve activity, a happy mood, and relaxation. In such a scenario, waterfalls become a significant wildlife resource for controlling visitors' moods by making them tranquil.

Despite the importance of waterfalls in wildlife tourism, it is still not defined and declared as a particular trip package in various travel itineraries. It happens as part of other tourism activities when wildlife visitors visit the waterfall as an attraction, not solely as a wellness package. In certain protected areas, the existence of thermal resources may provide an opportunity for health or medical tourism growth.

Most of the protected areas bisected by the rift valleys in East Africa have substantial thermal resources. Due to a lack of competence and political will, these resources are underused for thermal health tourism development in many protected locations. It is a type of health tourism that involves using subsurface hot or cold water to treat or prevent mental, joint, and inflammatory ailments in humans. In most circumstances, it is termed a thermal bath.

Therefore, therapeutic qualities, such as bathing in hot springs found in thermal resources, enhance blood flow, circulation, metabolism, and the absorption of important minerals. Furthermore, health tourism may entail drinking water from a variety of drug sources.

In Tanzania, drinking water from the River Mapacha (in the Swahili language Mapacha means twins) situated in the Tanganyika administrative district, might practically lead to the birth of twin infants. On several occasions, many women had proclaimed to have twin infants after drinking the spring water. However, incorporating such unpublished and scientifically disapproved traditional entities into medical tourism still needs scientific investigation and proof.

Citing an example that dates back between 2010 and 2012, when a traditional healer in the Samunge village, adjacent to the Ngorongoro Conservation Area in Tanzania, thousands of people from all walks of life in East Africa traveled to the late Mwasapile's residence in Samunge village in Loliondo, near Ngorongoro in northern Tanzania, where he distributed a liquid herbal medicine, collected from wild plants, that allegedly could virtually heal any chronic illness, including diabetes, cancer, and HIV/AIDS.

However, the occurrence of medical procedures is rarely viewed as either a form of medical tourism or wildlife tourism. At the destination and national levels, medical tourism hotspots and services are neither institutionalized nor incorporated into the tourist master plan. That makes the marketing of such places especially challenging since even infrastructural and superstructural improvements to enable medical tourism operations at the destination are pending.

Wildlife Filming Tourism

It is sometimes considered either phototourism or film-induced tourism. It is a form of wildlife tourism for photogenic visitors; wildlife tourists look for sites with beautiful surroundings and exceptional animal resources for favorable and easy filming.

Wildlife filmmaking is one of the fastest-developing specialties of wildlife tourism around the globe, including direct visitations to the protected areas associated with wildlife movies or to the production houses related to wildlife movies. There are many strategies a film tourist might employ to identify favorable areas for film tourism. Repeated broadcasting and airing at certain places may also attract more visitors.

It is the technique that allows wildlife tourists to go to natural locations to take pictures or videos for personal or business uses. The purpose of photography is to capture attractive wild places, natural flora, and animals. For a better recording of the wildlife film, wildlife tourists should have an in-depth awareness of wildlife ecology and behaviors. Such awareness may enable wildlife tourists to easily find wildlife species and recording locations.

Identifying and locating wildlife species in protected areas is not as simple as locating the same species in zoos or zoological gardens, where a visitor can always find an individual in the same position. Wild animals are mobile and situational; they change places and behaviors depending on the circumstances and availability of ecological resources. Wildlife freely wanders and occupies various sites at different times, as free-ranging animals do. In similar cases, for wildlife visitors to exactly spot the animals, they have to be aware of such behaviors or engage wildlife ethologists.

Wildlife shooting is completely adventurous in nature, which may sometimes compel filmmakers or photographers to waste a lot of energy accomplishing their aims. Similarly, photography often requires a long journey through numerous animal habitats, including woods, woodland, shrubland, desert, marshes, and grassland, among others.

Wildlife species occasionally occupy the same area throughout the year or their whole life cycle. Also, filming and photography often take a lengthy time, as the size of the wildlife species, the activity of the animal, and the kind of vegetation may delay the photographers and filmmakers from finding the right species.

In certain circumstances, it is safe to label wildlife photographic tourism as a dangerous recreational activity. The proximity of wildlife species while recording or taking pictures also justifies when it is dangerous. Several recorded and unverified occurrences exist where wildlife photographers and filmmakers have been either killed or critically wounded by wild animals. However, the danger of photography may vary depending on the goal, the behavior of the species targeted, and the type of protected area where wildlife filming takes place.

Video or photography of friendly wild animals, such as many birds, tortoises, and wildebeests, among others, require less energy, time, and hazards than filming aggressive animals with the correct timing and approach. Filmmakers have to invest in modern equipment to reduce time and energy when identifying and approaching closely dangerous wild animals such as the African buffalo, lion, African elephant, and leopard with zero injuries and fatalities to the filmmakers and wild animals.

Any minor error made by a wildlife photographer when filming or photographing, such as the detection of flashlights by a hostile wild animal, may lead to fatalities, injuries, and damage to property. As a result, wildlife photography is characterized as adventurous because wildlife photographers usually anticipate the unexpected when shooting and recording adventures.

As a freelance tour guide in 2006, I observed a tragic occurrence in one of the protected areas in the Southern Tourism Circuit of Tanzania. A wildlife photographer mistakenly and closely approached the group of African buffalo on foot without knowing what was behind him. Unfortunately, a charged bull of African buffalo was rejoining the group after chasing away the lion. His girlfriend, in a tourist van or car, saw the bull and suddenly shouted at it to scare it away and rescue her boyfriend without any success.

Though it was too late for the rescue, the tour guide and driver immediately drove the vehicle fast to block the bull from hitting the photographer but only managed to frighten the buffalo away after kicking the photographer at once. Sadly, another tourist in the automobile was busy picking up the picture of the kicked wildlife photographer, who was now swimming in the air following the kick.

After the rescue, the critically wounded and fortunate photographer was moaning solely about his left foot and back problems. He was transported to the hospital, and I did not know about his destiny.

The event provides several lessons for wildlife tourists to learn. Knowledge and voluntary compliance with protected area regulations are important to wildlife visitors, tour guides, tour operators, and other tourism stakeholders. I still recall Professor Alexander Songorwa, who, while talking about wildlife tourism, sometimes argued that tourist operations can sometimes be as dangerous as a gun, which, in the wrong hands, can destroy everything, including the shooter and everything around him. Can you imagine having a completely loaded rifle in the hands of a crazed person?

As a result, conservation agencies and location management have active tourism policies, rules, and regulations to oversee wildlife tourism activities and visitor behaviors. It is the responsibility of wildlife visitors and operators to follow destination legislation and laws as a means of preserving the integrity and welfare of wildlife tourists, the environment, wildlife, and residents. The majority of those restrictions explicitly state the right location, time, and process for wildlife photography and filming.

If things were conducted correctly and lawfully, the illustrated wildlife photographer in the preceding lines would not be engaged in such hazardous activities. Such an error happened due to the recklessness of the tour leader and all the animal guests in the vehicle. Since it was a guided tour, the guide would have informed all the visitors of the dos and don'ts in the protected areas before the trip. In summary, this is a basic summary of what might happen if a wildlife photographer or videographer screws up with hostile species in the environment.

Moreover, photographs and video production take time and resources. As a patience-testing wildlife leisure activity, the photography of filming may take a long time, depending on the purpose of the activity. Commercial wildlife shooting demands attractive scenery, distinctive exhibitions, and the interaction of animal activities. The video of the display and interactions may take months or years to film.

When visitors demand a film of mating wild animals, the photographer needs to have patience until the breeding season comes. For example, the synchronized birth of migrating wildebeests in Serengeti National Park occurs in November. The same applies to those who are interested in recording or photographing the breeding of lions to catch the honeymooning of the lions that occurs in particular months of the year. Interested wildlife photographers and filmmakers would not want to miss the two stated ecological phenomena, but their timing or patience are equally essential.

Scuba Diving

It entails diving underwater to investigate inaccessible animal resources using specific devices. It is usually done in aquatic protected areas to observe animal resources. It is fairly widespread in Greece and Costa Rica. In every dive, wildlife visitors have the opportunity to discover and appreciate the beauty and outstanding endowments in the aquatic environment. It is also regarded as an adventurous aquatic activity containing limitless hazards and wonderful moments.

Scuba stands for Self-Contained Underwater Breathing Apparatus. It is from this acronym that underwater wildlife tourist activities acquired their name, since, in aquatic environments, divers cannot breathe without the support of such devices. In other words, the need for such devices lengthens the period of stay underwater to improve the performance of wildlife tourist activities.

Scuba diving provides visitors with a variety of experiences because it offers a range of activities and interpretations for wildlife tourists. Some wildlife visitors think of scuba diving as an adventure and sport tourism activity. Apart from admiring the beautiful underwater environment and species, some wildlife tourists are so hooked on the activity that they consider it their life. The energy invested in scuba diving makes wildlife visitors view the action as a sporting effort.

More significantly, scuba diving also sustains some contemplative qualities; that is to say, scuba diving has both physical and mental value for wildlife visitors. It is sometimes referred to as hydromeditation or underwater meditation because it promotes physical and mental relaxation.

The following features make scuba diving a useful contemplative tool for wildlife travelers: Scuba diving is a significant mental meditation technique, as it exposes wildlife tourists to the other side of the planet Earth, where they engage with a wonderful underwater environment and wildlife species that are entirely different from the ones they are accustomed to in the terrestrial environment.

In the process of satisfaction and appreciation, the wildlife tourist concentrates totally on undersea endowment while forgetting most of their everyday concerns and enhancing contentment. The thrill of hearing and witnessing aquatic species makes wildlife visitors feel like they are in separate worlds.

When using the breathing device underwater, wildlife visitors breathe slowly and deeply, which is an important aspect of meditation. Such a form of breathing assists the nervous system to relax and slow down the heart rate and blood pressure. In this manner, the wildlife tourists decreased their worries and enhanced their happiness, making them physically and mentally calm.

The underwater natural environment is rich in natural color, lights, and noises from varied aquatic creatures, currents, and waves, which are not only significant as attractions but also as psychological therapy. Studies claim that the bluest color of the water may ease the mind of wildlife tourists, while the sunshine is responsible for the stimulation of serotonin in the body of wildlife tourists, which is medically responsible for the regulation of human mood.

Scuba diving as a pure leisure activity does not just entail admiration of aquatic species but also diving underwater to observe physical features such as caverns and buried wrecks. As recreational divers, scuba enables wildlife tourists to dive into sunken wrecks and caves to carefully inspect their morphological traits and engage with wildlife species residing in the caverns.

In such circumstances, wildlife visitors feel unlike any other trip experience on planet Earth. However, scuba diving, as an exciting form of wildlife tourism, is not free from limitless hazards; some of them include dangerous encounters that usually result in deaths and injuries.

Mountain Climbing

It is a form of wildlife tourism that includes ascending to the highest attitudinal locations for socio-cultural and tourist activities. Mountain climbing, also known as mountaineering, is one of the most common activities in mountainous protected areas where relatively high mountains, hills, and valleys, such as Kilimanjaro and Arusha National Parks in Tanzania, attract wildlife tourists for climbing Mount Kilimanjaro and Mount Meru, respectively.

Mountaineering occurs in relatively restricted and constrained geographical areas. Such locations have outstanding scenic, climatic, topographic, and faunal qualities. Like other forms of tourism, climbing may equally contribute to socio-economic development initiatives at the location.

Moreover, the mountain may also be a tourist destination by itself, since the picturesque aspects of any mountain may attract animal visitors. In the same vein, the duration and presence of snow on Mount Kilimanjaro pique the interest of many wildlife visitors. That is to say, wildlife places may separately promote and offer mountaineering as a standalone trip package to wildlife visitors or integrate it with other tourism activities.

Mountaineering is frequently described as an adventure because it entails many unpredicted dangers for visitors and animals. Tour companies and guides frequently advise guests about the potential results before ascending the mountains. Travelers may also require unique mountain climbing tools to effortlessly scale the mountains. The traveler may have their own climbing equipment or lease the same gear from a tour company or any other source.

Additionally, climbing is a protracted tour activity that may last up to several weeks; it exposes wildlife visitors to numerous climatic zones and ecological places before, after, and during the summit. Mountaineering may be independent or assisted, where, in particular regions or conditions, wildlife visitors need specific assistance to easily ascend the tallest mountains.

Tour companies or location management may provide extra assistance to the summit, such as helicopters and cable cars, among other things. Destination management and tour providers may advise wildlife guests in advance on where and when to rent climbing gear and other associated equipment.

For wildlife visitors, mountaineering activities involving more than 1 day may require supplementary services, such as housing and culinary services. In most situations, tour operators are also involved in providing tented camps and culinary services to aid mountain climbing; this may also include carrying personal goods like bags when climbing. Travel operators and protected regions are also responsible for provisioning emerging basic first aid and rescue services.

It is one of the most physically and emotionally demanding forms of wildlife tourism, as it largely involves mental preparation, enough physical strength, extraordinary resilience, endurance, body sharpness, body perfection, and flexibility.

During climbing, wildlife tourists may anticipate the unexpected; among others, the occurrences of fatalities and acquisitions of physical, physiological, and psychological issues for wildlife tourists during and after mountaineering are equally likely.

Because of the rapid attitude shift, it is normal to observe wildlife visitors dying and experiencing heart and lung difficulties during and after the summit. During climbing and descending, wildlife visitors may encounter unpleasant changes with variable atmospheric air pressure, adversely interfering with their breathing systems. Depending on the capabilities and convenience of first aid and rescue crews, some wildlife visitors end up dying uphill. Though mountain climbing might be fatal, it is still exciting.

Despite all the criticism, climbing continues to be one of the most fun and exciting nature activities, notably for adventure visitors. It might be the opportunities for adventures to increase their endurance, body toughness, and flexibility, as well as their wilderness survival and life abilities. It is unique in the search for personal identity, as mountain climbers continually aspire to beat the global climbing records. Mountaineering is not only about summiting but also about amassing experience for future challenges.

Additionally, mountain climbing connects wildlife visitors with a diverse natural environment. Due to the proximity of the high mountains to the sky, it makes wildlife visitors feel like flying, standing on the top of the globe, and touching the sky, which is surely a life-enriching experience. Since it requires a lot of walking, it puts wildlife visitors near natural vegetation and wild creatures of all varieties.

Let's start with the basic example of Mount Kilimanjaro, from Kilimanjaro National Park. The visitors appreciate climbing it since, by mounting the mountain, they have gone from tropical to temperate climatic areas. That simply implies that Mount Kilimanjaro contains both temperate and tropical climatic characteristics. In other words, by climbing Mount Kilimanjaro, visitors may witness and experience both temperate and tropical wild flora and animals.

Moreover, the mountains are sites of spiritual or cultural value; presumably, they are the point where the earth and heaven meet. The summits of mountains are seen as the link between earth and heaven. In many countries, mountains are holy sites where wildlife visitors may discover the cultures and religions of diverse tribes. Various mythologies and religious expressions of the mountains make them spiritually important.

In conclusion, mountains play a vital role in religious and spiritual values. In Christianity, mountains have settings for various key historic religious events, such as Moses receiving the 10 commandments on Mount Sinai. In Judaism, mountains signify divinity; Mount Moriah is where Abraham sought to sacrifice his son Isaac. Also, in Islam, mountains as a part of nature signify the vastness and strength of Allah, since Mount Hira is a sacred spot where Islam accepts that prophet Mohammad SAW got his first revelation.

In Hinduism, the Himalayas view Mount Kailash as holy to Lord Shiva. In traditional religions, mountains are viewed as the property of gods, indicating perpetual life and a powerful link to the gods because mountains are immovable, transcendent, unshakeable, and unreachable, exactly like their gods.

In traditional communities, mountains are the particular and metaphorical destinations for spiritual pilgrimage. In those cases, wildlife tourists have a lot to learn and practice from mountaineering. Wildlife tourists can climb the various elevated points with the protected areas to enjoy the uphill biodiversity, learn about the spiritual aspects of the mountains, pay a visit to the holy places found in the elevated points for worship and religious learning, and also consolidate their mountain climbing experience.

Canoeing Tourism

This is a type of tourism in which a boat is used to discover cultural and natural landscapes along or inside water bodies such as lakes, rivers, wetlands, or the ocean. It is also called kayaking; it is commonly practiced in different kinds of protected areas to reach locations that may not be accessible by surface and air transport. It is common in the island's national parks, wildlife reserves, and marine protected zones.

In this type of wildlife tourism, visitors go by canoe or kayak that is operated and owned by either tour companies or tourists themselves. It is a multifunctional and multidimensional wildlife tourism activity for creating adventure, cultural, sporting, and leisure experiences.

The choices and aims of the wildlife visitors always establish their purposes and ambitions for kayaking. It is possible for wildlife tourists to integrate all of their goals into one trip, striving for a variety of experiences on either one or numerous journeys. In other words, canoeing operators may develop canoeing in different ways to fit the tastes of varied wildlife visitors.

Canoeing is a kind of tourism that includes traveling by canoe or kayak while studying natural or scenic characteristics at the destination. It can be a leisure activity, a sport, an adventure, or a cultural experience, depending on the wildlife tourists' visiting goals.

Additionally, canoeing tourism provides several advantages to wildlife visitors, such as increasing physical and mental health, lowering stress, developing environmental knowledge, and helping local communities. It may take place in public, private, or community-protected areas, or outside the protected environment.

Moreover, canoeing may take place in unprotected wetlands, rivers, lakes, and even the ocean. It is difficult to find dedicated canoeing operators in undeveloped countries; as a result, many wildlife tour companies include canoeing as part of other wildlife tourism packages. However, in the developing world, specialized and competent firms provide canoeing activities as a standalone package.

On numerous Norwegian islands, it is possible to explore and enjoy the landscapes as well as observe mountains, glaciers, and animals by boat. Canoeing provides domestic and international visitors in Brazil with the opportunity to discover and enjoy the rich biodiversity and indigenous culture of the jungles surrounding the Amazon River. The same applies to kayaking across the Galapagos Islands in Ecuador and the Zanzibar Islands in Tanzania.

Furthermore, the kayak provides a greater opportunity for wildlife visitors to enjoy the area's culture, nature, and picturesque elements. The use of houseboats while visiting the region makes kayaking a unique and fantastic wildlife tourism activity. The houseboat gives memorable experiences by immersing wildlife visitors in real rural life and the cuisine of the location.

Furthermore, the kayak provides a greater opportunity for wildlife visitors to enjoy the area's culture, nature, and picturesque elements. The use of houseboats while visiting the region makes kayaking a unique and fantastic wildlife tourism activity. The houseboat gives memorable experiences by immersing wildlife visitors in real rural life and the cuisine of the location.

The houseboats, as floating restaurants or bars, provide a great eating and drinking experience for wildlife visitors with maximum protection and privacy. To satisfy the diverse expectations of wildlife guests, the operators develop numerous kayaking programs. Engagements, birthdays, and honeymoons are among the rising tourism activities in kayaking.

Walking Safari

A walking safari is a guided or unguided excursion that requires wildlife visitors to walk while observing wildlife resources. Instead of using an automobile or tourist van, wildlife tourists walk inside or outside protected areas to appreciate the wildlife resources. A walking safari is also called a pacing safari or safari, as it includes the study of the natural environment on foot. It is a form of wildlife-watching activity that scarcely depends on vehicle protection from aggressive creatures. Walking safaris may be as safe as game drives, provided all wilderness measures are taken into account.

During a walking safari adventure, wildlife visitors may encounter both peaceful and threatening wild creatures. However, walking safaris continue to be an energy-demanding, life-threatening, and costly form of wildlife tourism activity in certain wildlife tourist locations. The demand for competent and professional tour guides is inevitable. Throughout the walking safari, the guides, who are competent in wilderness survival skills, may deliver an unquestionable wilderness experience to wildlife visitors.

It is often undertaken in terrestrially protected areas, in places with suitable plant types to permit easy observation of animals. Woodland, shrubland, and grassland vegetation may be important for a walking safari. Forests and forest plants may hamper a walking safari. In all terrestrially protected zones, walking or pacing tourism is practiced.

A walking safari provides a more exceptional experience than other types of wildlife tourism because it takes wildlife visitors closer to nature. Wildlife tourists' journeys through the wilderness may teach them a lot of things, from wilderness survival skills to nature. It gives a chance for wildlife visitors to also study the habits of wild plants and animals, an opportunity that most wildlife tourists may miss on game drives.

Through a walking safari, wildlife visitors may discover the flying distances of different wildlife species, animal trails, vocalizations, and their physical traits. With competent and bilingual tour guides, wildlife visitors acquire the identifying abilities of numerous wild plants and animals without or with minimal disruptions to nature. As it includes the identification of insects, birds, and animals, a walking safari works as a volunteer learning platform for wildlife visitors.

Walking safaris may be highly challenging, particularly for wildlife travelers, those with physical limitations, the elderly, and children. As it includes long and tough hikes, walking safaris naturally become not everyone's cup of tea. It is important to note that not every older, handicapped person or youngster can walk, but it may be difficult for the majority of such wildlife visitors to deal with problems associated with walking safaris, including learning outdoor survival skills.

The obstacles of a walking safari include trekking through challenging terrain for an extended period, but others are brief and last for a few hours. A walking safari is also an adventure because it involves meeting the unexpected. The presence of many types of wild plants and animals may accidentally create severe health concerns for wildlife visitors. For example, wildlife visitors who are allergic to insect bites, spores, and plants may suffer certain symptoms following physical contact.

However, pacing or walking Safaris are amazing wildlife tourism experiences that give wildlife visitors the ability to see a diversity of wildlife species, including the big five. It becomes intriguing to carefully see lions, leopards, elephants, rhinos, and buffalo grazing, breeding, and foraging in their natural settings. It becomes even more intriguing to see such exchanges on foot in a calm setting.

The walking safari is quite widespread in the African wilderness, in public, community, and private protected areas. Also, a walking safari might take place in an open or unprotected environment. Walking safaris also provide wildlife visitors with a chance to meet with residents and understand their culture in unprotected territory or community-protected areas. Interaction with culture gives extra experience to wildlife visitors and helps them better grasp the links between wildlife and traditional knowledge.

Africa contains exceptional wildlife tourism sites for walking safaris. Zambia is one of the unquestionable places where South Luangwa and Kafue National Parks offer a fantastic walking safari experience to view elephants and giraffes. Giraffes, zebras, and hundreds of bird species. Rwanda is always fantastic for viewing mountain gorillas on foot in their protected environment.

Volcanoes National Park is one of Rwanda's most well-known protected landscapes for gorilla tours. Furthermore, Kenya is one of Africa's most prominent and oldest walking safari locations. Masai Mara National Reserve, Amboseli National Park, and Tsavo National Parks are famed for providing superb walking safaris on the African continent. In southern Africa, Namibia provides a blend of wildlife and physical elements in its walking safari programs. In a walking safari adventure, wildlife visitors witness animals, dunes, Rocky Mountains, and desert environments.

The main thing to notice is that involvement in walking safaris is not free; it requires the wildlife visitor to pay an admission fee, and more crucially, it also requires the walking safari operator to pay for walking safari operating licenses. However, the operation license and entry costs may vary according to the nation, tourism season, and category of protected sites.

Game Viewing

It is sometimes called wildlife watching, wildlife gazing, or game driving. It entails observing wildlife in its natural surroundings from the car. It entails studying animals while eating, mating, fighting, strolling, or hunting. It also incorporates observation of plants, lakes, seas, rivers, rocks, or mountains in the protected regions while driving.

It is one of the most basic and oldest forms of wildlife tourism activities. In most situations, it is not confined to seeing wildlife while driving; it also extends to whale watching, bird watching, lion watching, butterfly watching, mountain climbing, and boat safaris. It serves as the main source of animal tourism and the foundation for further pursuits like hot air balloon safaris, walking safaris, and canopy trekking.

Game watching is a characteristic and important wildlife tourist activity that takes place in practically all kinds of protected areas, whether public, private, communal, or open. Certain wildlife tourism activities can only take place in particular protected regions due to legal, geographical, and policy effects. As previously stated, trophy hunting is a lawful activity in Tanzania, but it is strictly forbidden in all of Tanzania's national parks.

Hunting of any sort is forbidden in Kenya but permitted in Botswana and South Africa. Canoeing is only possible in aquatic protected zones, but it is impractical in terrestrial protected areas. Since it is an entirely non-consumptive tourism activity, several nations and protected regions accept game-watching as a relatively safe tourist activity.

Wildlife visitors can observe wildlife in a variety of ways, such as through guided or private trips, game drives, or strolling. The descriptions of wildlife-watching activities vary depending on the location or the equipment used. Animals from above the earth are referred to as hot air balloon safaris, whereas game drives are referred to as cars.

Additionally, observing animals while walking is termed a walking safari, and viewing the aquatic environment using different aquatic equipment is called kayaking, snorkeling, or scuba diving. Based on the particular kind of wildlife species, the observation of bird species is referred to as birdwatching. It is one of the most prominent game-observing activities around the globe.

Game-watching may be a fascinating and informative experience, as you can observe a variety of species, such as lions, elephants, giraffes, zebras, rhinos, and leopards, and learn about their behavior, ecology, and conservation. It is generally secure and pleasant because it attracts wildlife visitors of a variety of ages, physical conditions, sexes, and budgets.

Furthermore, wildlife visitors may engage in game-watching on packaged or non-packaged trips. International travelers frequently engage in wildlife tourism via tour companies. A tour operator offers a bundled trip that includes local transport, hotels, food, drinks, guides, and admission costs to animal sites for travelers.

Many domestic visitors may plan their hotel, local transport, and culinary services. Only a few domestic visitors hire tour operators for their trip owing to familiarity with the surroundings and restricted cash. However, domestic visitors may also require additional services from animal attractions, such as tour guiding and unique touring instruments.

Cycling Wildlife Tourism

Traditional wildlife tourism focuses on observing animal resources while in a tourist van. Many wildlife visitors and management believe that the most popular means of experiencing wildlife is via game drive. It has been the case for many years in developing nations, notably in Africa south of the Sahara, where wildlife tourism forms the backbone of tourism.

In short, wildlife tourism encompasses a wide range of activities, including bicycle wildlife tourism. Just like the conventional game drive, where visitors study wildlife resources while driving, bicycle tourism also entails observation of wildlife resources while riding either independently or in groups. It is another way to appreciate the outstanding animal riches in their natural surroundings.

Riding wildlife tourism is a fully terrestrial experience that exposes wildlife visitors to several dangers when riding. As wildlife visitors view the natural resources without the protection of cars, despite all the concerns described, wildlife cycling tourism is as safe as game driving, owing to the preventative precautions that relevant authorities and operators normally take before, during, and after the ride.

It is a common type of wildlife tourism activity in industrialized countries, provided by expert wildlife tour operators. A cycling wildlife tourism package may encompass a single or numerous protected areas. Throughout the excursion, the wildlife tour operator may be responsible for providing bikes, cycling equipment, food, hotel facilities, and cars for guiding, as well as the possibility of rescue services. Special or certified bike equipment and training are available in Wildlife Cycling Tourism to minimize any unwanted delays and mishaps.

Additionally, wildlife cycling tourism may encompass one or more countries in a single or separate vacation package. It may also require more than one wildlife operator to perform the same itinerary. In wildlife cycling adventures, visitors view a broad variety of animal resources, including zebras, lions, cheetahs, and numerous birds, among others.

Depending on the wildlife cycling location, tourists may come into contact with local people and culture while riding at their destination. When riding in communal and private protected areas, wildlife visitors may physically connect with local populations and learn since wildlife tourists can't interact with local people in the protected regions where human presence is restricted.

Moreover, wildlife cycling tourism comprises a self-guided trip package. Depending on the policy of protected areas and the location, wildlife visitors may be able to participate in self-guided wildlife bicycle trips inside and outside protected areas, depending on the policy and location. As an individual package, the wildlife cycling tourists plan most of the riding supplies, food, and hotel themselves before the cycling excursion.

In summary, wildlife cycling tourism delivers physical well-being, cultural experience, and natural admiration. As a result, various wildlife attractions in Africa are steadily adding bike safaris to diversify wildlife tourist activities. One of the frequent sorts of wildlife cycling safaris is the Himalayas Adventure Trust, which normally arranges cycling safaris for wildlife visitors to explore and enjoy the sights of the Himalayas in Nepal, India, and Tibet.

Treasure Hunting

Treasure hunting is the sort of wildlife tourism activity that requires wildlife tourists to follow the supplied set of directions to specified spots in the protected areas to discover concealed objects. In this challenge or activity, the treasure is purposefully buried or hidden without the knowledge of wildlife visitors so they can find it within the stipulated time. In the treasure hunt, the particular management zone of the protected areas becomes the playing surface or ground for wildlife visitors.

Treasure-seeking remains an enjoyable and participatory activity for wildlife travelers looking to discover the peculiarities of certain protected regions. It provides wildlife visitors with the opportunity to learn about the specific protected areas and the resources linked to them, making it both enjoyable and instructive. Rich seekers have the chance to share their life experience, expertise, and talents while they strive to uncover the buried riches.

In treasure hunting, groups of wildlife visitors wander throughout the protected areas to overcome carefully designed obstacles and eventually gain points. As they stroll through the protected region and attempt to fulfill the task, wildlife visitors enjoy engaging in wildness, with multiple pauses in each beautiful location.

Treasure hunting is an informal walk or self-guided trip to protected and unprotected regions. The operators of treasure hunting give pleasant recommendations to wildlife visitors to uncover the hidden jewels without personally taking the tourists to the place. In such tasks, the prize is anything that can readily be traced and seen.

The challenges occur in protected regions dominated by peaceful wild animals or in places with a few dangerous wild animals. The treasure hunters are typically briefed on basic wilderness survival skills before the trip as a way of safeguarding tourists against any hazardous encounters during the obstacles. The hidden treasure may be given either a trophy, certificate, or any unique present upon successful discovery.

It is one of the promising wildlife tourist activities in Africa, where many protected areas still sustain traditional wildlife tourism activities. Some public, private, and community-protected areas contain magnificent terrain that may also assist in treasure digging. That is to say, the protected regions with limited species diversity and richness are logically relevant to treasure-seeking activities.

Saanane Island National Park, situated inside Mwanza City in Tanzania, is an example of how adopting treasure-seeking as an extra wildlife tourist activity is viable. It is because the park lacks aggressive wild animals and has favorable flora and topography for treasure hunting. In summary, many wildlife attractions may design and include treasure-seeking as one of their wildlife tourist activities.

Canopy Walking

A canopy walkway is an aerial tourist activity that includes going across the higher areas of the site or forest on a hanging bridge to view animal resources. The canopy walkway encourages wildlife visitors to carefully observe animal resources in the natural environment from an aerial view.

As a solely wildlife tourist adventure activity, a canopy walkway is an aerial wildlife tourist activity that allows wildlife visitors to obtain an aerial perspective of wildlife resources while walking on artificially and traditionally made hanging bridges. Canopy walking may fit in both ecotourism and adventure alternative kinds of tourism, as it entails going to a natural environment in small groups to participate in vigorous exercise. It entails walking on a suspension bridge on top of a rainforest.

The suspended bridge is normally constructed of wooden or metal materials and ropes, but it has been scientifically designed to be safe and secure. In most instances, the bridges are fashioned of locally available materials to obscure and disguise the natural landscape. When creating such bridges, always put the safety and security of wildlife visitors first.

Kakum National Park, Ghana, was one of the first African national parks to introduce the Canopy Walkway as a tourist attraction. In East Africa, Udzungwa Mountains National Park, Tanzania, was the first wildlife destination to create a canopy walkway.

Through the pathway, nature tourists may encounter melodic birds, mountainous vegetation, animals, and insects that live on the roofs of protected regions. The hanging bridges provide wildlife travelers with unique views of animal resources that are technically unseen from the ground.

The canopy walkway is normally managed by the respective management authority of the protected area but in compliance with each protected area's general management plan and rules. The GMP and regulations regulate the regions, seasons, and processes for carrying out canopy walkway activity.

Apart from establishing areas, charges, permissions, and schedules for walkway activity, the GMP and regulations offer safeguards for such vulnerable environments and enhance both the security and safety of wildlife tourists while boosting the delight of the visitors while walking expeditions.

Furthermore, the canopy walkways provide instructive and unique experiences to wildlife visitors, since most tourists feel like strolling on top of natural trees or treetops. Wildlife visitors learn about the canopy, arboreal biodiversity, and meteorological conditions.

Hot Air Ballooning

Hot air ballooning is an alternate kind of airborne adventure, allowing wildlife visitors to explore wildlife resources while soaring on a hot air balloon. It is an alternative option for travelers looking to explore nature's resources from above. That implies that a wildlife tourist has numerous options for viewing animal resources. Alternatively, wildlife visitors may employ many means of travel, including road, train, air, and sea, to explore animal resources.

It is critical for sustainable wildlife tourism because it provides a unique experience and limits physical encounters between animals and visitors. Several wildlife tourism sites provide trips on hot-air balloons but at an expensive fee. The balloon is made of nylon-like materials that are inflated with hot air using propane burners. It also has a special cabin for travelers to stay in.

It operates using a simple tactic: the hot air inside the balloon, burnt by propane burners, makes it less dense than the surrounding air; therefore, it lifts the whole structure off the ground surface, enabling it to float in the air smoothly at the appropriate height where tourists can have a sufficient view of the wildlife resources with or without significant disturbances to wildlife.

It is always safe to travel in a hot air balloon since the pilots are normally exposed to regular training, and adherence to tight flying standards and regulations is part of their training. Registration of each balloon's equipment and area of operation is necessary for several nations. The owners normally follow regular maintenance and inspections as part of professional and legal norms.

The hot air balloons may similarly be used for scientific study, inquiry, habitat monitoring, and teaching objectives, especially in difficult terrain. Wildlife tourism managers may use the balloons as a platform for monitoring and analyzing atmospheric conditions.

It may also be used as a platform for monitoring wildlife tourism activities, allowing tourism managers to map diverse wildlife resources to understand their exploitation. The overhead perspective provided by hot-air balloons is critical for aerial image capture and quick monitoring of animal resource status.

It is usual for some individuals to use a charter balloon for celebrating different social occasions, including birthdays, reunions, weddings, anniversaries, and even business gatherings. Rejoicing on significant occasions in the charter balloons sounds like rejoicing while soaring without wings. Furthermore, when it takes place in natural destinations, the enjoyment of animal resources takes precedence over the festivities, making the balloon ride a memorable experience for the guests.

Despite the various advantages that hot air ballooning gives to wildlife tourists, if the environmental and socio-economic repercussions of the balloons are neglected, their imprints will taint the sustainability efforts of sustainable wildlife tourism. As an example, the utilization of hot air balloons for festivals is prevalent in developed nations; the deployment of balloons for enhancing festivities and entertaining unity among communities within wildlife-protected areas should be avoided to reduce disruption to animals and visitors. Allowing balloon festivals in the protected landscapes might affect the protected regions in some ways since they always take glows, races, and parades as part of the festivals.

Consumptive Wildlife Tourism Utilizations

It is a form of wildlife tourism that entails the complete offtake of animal resources from their native locations. It is consumptive in the sense that it involves lawfully but physically off-taking of wildlife resources from protected and unprotected landscapes. It generally occurs in publicly, communally, or privately managed protected landscapes. The most common form of consumptive wildlife tourism is trophy hunting.

This type of wildlife tourism has sparked unending arguments from a wide range of wildlife tourism players, including animal advocacy groups, trophy hunters, environmentalists, and non-governmental organizations. Animal advocates have consistently described consumptive wildlife tourism as a cruel activity, immoral, and a key source of species extinction and habitat destruction worldwide.

On the other edge of the discussion, certain protected areas, local communities, and wildlife tour operators deliberate how to leverage the advantages of consumptive forms of wildlife for environmental protection and socio-economic development. In light of this, it is vital to state that the discussion on the need for consumptive forms of wildlife tourism is complicated technically, politically, socially, ecologically, and legally.

It demands considerable attention and insight to balance the social, political, environmental, and cultural worlds. With this consideration, the interests of numerous wildlife tourism stakeholders can be taken into account. Some of the most popular types of wildlife tourism include trophy hunting, resident hunting, therapeutic hunting, fisheries tourism, and pleasure hunting.

The History of Trophy Hunting

Tourist or trophy hunting is consumptive wildlife exploitation that takes place within or outside private or public protected areas for trophies and leisure pursuits. It is one of the most well-known tourist initiatives worldwide, and it includes the slaughter of animals for recreational purposes. In essence, hunters gather trophies for a variety of purposes, including decoration and traditional remedies.

Trophies are in numerous forms, such as animal skulls, skins, ivories, horns, and tails from major sports species like lions, leopards, buffalo, elephants, and rhinoceroses for traditional rites. In Africa, trophy hunting originated during pre-colonial times, when traditional hunters sought game animals to collect customary-valued trophies. In pre-colonial Africa, trophy hunting happened on community territory for traditional rites.

In that period, trophy hunting imposed relatively trifling consequences on wildlife species owing to ingenious hunting weapons, few hunted animals, well-established and accepted customary laws, and a lack of international commerce on trophies. Later, a colonial government in Africa reformed trophy hunting techniques by developing and enforcing hunting rules that established and detailed hunting zones, game species, and hunting procedures.

The government modified the trophy hunting sector to reflect Western culture, which excluded traditional hunters due to strict laws and higher hunting costs. Furthermore, numerous regions with tremendous wildlife variety were taken by colonial administrations, delineated, and gazetted as public game reserves.

Certain historic gazetted places, such as Selous Game Reserve in Tanzania, are still administered as game reserves and national parks in various regions of Africa. African traditional hunters were eventually alienated from wildlife and exploitation because they couldn't afford to pay for game hunting fees and permits.

Game hunting was vital under colonial regimes because their lifestyles and economies relied on skins, ivories, manes, and other valued trophies—not only as raw materials for industrialization but also as money to spend in the barter trade. Soon after the introduction of Western hunting techniques, trophy hunting began producing a dramatic decrease in species variety.

It is because, in the beginning, the colonial government considered contemporary hunting equipment and vast gazettements of game reserves to be the primary scientific criteria to increase sustainable tourist hunting.

Trophy or tourist hunting was common in Africa in the 1970s; this was the period when wildlife conservation around the globe started to encounter major environmental repercussions from trophy hunting (Cruise, 2019). To make the record accurate, this was also the period when the bulk of African nations had already won their independence but were still enforcing colonial trophy-hunting rules.

In the late 1970s, the consequences of trophy hunting amplified exponentially due to enormous commercial poaching of keystone species, in particular African elephants (*Loxodonta Africana*), black rhinoceros (*Diceros bicornis*), and lions

(*Panthera leo*), and it was difficult to technically differentiate between the adverse impacts of trophy hunting and commercial poaching.

In the early 1980s, environmental conservation problems began to gain momentum. It occurred during a period when a global movement on animal rights was gathering pace. Some animal activist organizations partnered with environmental conservation groups to stop cruelty to both domesticated and wild animals by openly denouncing tourist hunting and laboratory experiments with animals.

Currently, there is a major discussion on whether to continue or discontinue tourist hunting. Some conservation stakeholders believe that hunting is harsh on animals and pushes certain species to extinction. The biggest proponents of this argument are animal and environmental campaigners, while the opponents are trophy hunters.

However, there are also different thoughts among environmentalists and tourists regarding trophy hunting, as some of them passionately favor it while others frantically condemn it. According to opponents, trophy hunting is ecologically and socio-commercially hostile, while proponents portray it as an environmentally constructive instrument and an economically profitable entity.

Each debate side strives to explain the validity of its arguments by offering numerous supporting pieces of evidence. In brief, the discussion revolves around animal care, rights, and usefulness. Of all, the most contested subject is the sustainability and extinction of wildlife resources, with particular implications for the ecosystem, conservation status of the hunted species, and socio-economic integrity of the hunting locations.

Tourist hunting grounds are in government-, municipal-, or private-protected areas. Trophy hunting technique varies across nations due to differences in animal variety, conservation policies, climate, hunting seasons, culture, and hunting legislation. The US generates the majority of trophy hunters in the world and is the leading importer of hunted trophies internationally, importing 10 times more than China, which is the second importer.

Some hunting sites become more renowned than others. Eastern and southern Africa are the most well-known hunting areas. Despite the diversity of hunting grounds in many hunting features and processes, the impact of trophy hunting on the hunting destinations may be the same.

Trophy Hunting Procedure

It is a type of wildlife tourism activity that aims to collect wild animal trophies. In this sort of consumptive wildlife consumption, the trophies are acquired via permitted hunting. To properly comprehend the meaning of trophy hunting, the words trophy and licensed hunting require a more extensive explanation.

Trophy typically refers to any portion of a hunted wildlife species chased by trophy hunters during a legal hunting excursion. As previously stated, an animal's trophy may be a dead or living animal or its components, including its head, horns, skins, eggs, features, flesh, bones, cranium, manes, tusks, hooves, and many more.

Therefore, trophy hunters search for trophies throughout the hunting journey. The type, size, and color of trophies may vary between and among game species. That is to say, a trophy hunter may target certain species because he cannot find the same kind of trophy from other game species.

For instance, the lion, *Panthera leo*, has highly formed manes that elephants do not have. The African elephant, *Loxodonta africana*, has tusks that lions don't have. The trophy hunter wanting manes will opt for lions, while the ones seeking fully grown tusks will go for African elephants.

Also, prize color differs across species; the skin of the zebra, *Equus buchelli*, and African buffalo, *Synceras caffer*, vary substantially, where the skin or fur of the African buffalo is more or less black while the zebra's skin has black and white stripes. Honestly speaking, I am a colorblind person who cannot determine whether a zebra is black with white stripes or white with black stripes.

Additionally, the prize colors may vary even among individuals of the same game species due to, among other variables, the kind of habitat the animal is residing in and the physical and physiological characteristics of that person. This leads us to endless scientific debates between unmanned and manned lions, the African savannah elephant, and the African forest elephant, to name a few.

In the previous paragraph, as a short and specific illustration, I mentioned manes as one of the trophies of African lions. A study in Serengeti National Park, Tanzania, and Tsavo National Park, Kenya, suggests that dark manes are a sign of a higher fitness level of lions. The darker lion has higher reproductive success, reproductive life span, and offspring survival than the lion without manes, according to the studies.

To conclude the dispute, it is vital to note that the same game species may have multiple colors of the same prize based on several conditions. Other research reveals that certain lions limit the size of their manes to prevent extra warmth generated by manes and also impediments created by manes while navigating around thorn bushes.

Another, possibly unclear or unfamiliar word, "hunted wildlife species," was added to the definition of trophy hunting. In the opening portion of this book, an explicit explanation has been given to differentiate between hunted and non-hunted animals. The beginning portion of this book clarifies further by categorizing all hunted animals as game species.

In other words, not all wildlife species are evaluated and included in trophy hunting safaris due to numerous factors, including the lack of appealing and big trophies and the conservation status of the wildlife species. Any animal species identified locally and globally as critically endangered species are excluded from trophy hunting packages owing to scientific and legal limits.

For example, putting black rhinoceroses or wild dogs in the hunting quota may prolong the species' extinction trend due to its low population and reproduction rate. Also, international agreements such as the Convention on International Trade in Rare Animals (CITES) forbid trophy hunting of rare animals as a strategy for saving endangered wildlife species from needless extinction caused by humans. The limits of CITES and other international accords are only applicable to the nations that have formally accepted the agreements.

Trophy hunting is also known as licensed trophy hunting, or legitimate hunting, in addition to the trophy keyword. To partake in a trophy hunting safari, a trophy hunter or tour hunting organization must receive a valid hunting license from a competent government. Hunting any game species, wherever there is a valid hunting license, may result in criminal prosecution.

However, not all game species need a hunting license, as certain governments may legalize the shooting of problem species, or "vermin," or invading alien species, as one of their tactics for lowering or eliminating their unmanageable numbers. Trophy hunters participate in the physical management of animal populations and the elimination of foreign-invading species. Such animals frequently lack natural predators to manage their populations.

The absence of proper management methods may boost the population size and status of foreign invasive species, making them the most devastating pests. Invasive species undermine ecosystem health, damage crops, and carry illnesses to cattle, people, and animals. For instance, wild pigs (*Sus scrofa*), an invasive species in the United States, generate over one billion dollars in agricultural losses. As a result, the US government uses deadly techniques, such as trophy hunting or recreational hunting, to limit its numbers.

When the proliferation of invasive species surpasses environmental carrying capacity, it affects the ecosystem itself, lives, and assets. In such instances, conservation authorities grant hunting quotas to hunting corporations to limit their numbers. It is also beneficial to remove extra people and harmful invading species from the reserves. Moreover, trophy hunting lowers the number of aged and sterile animals, which are commercially, socially, and ecologically unproductive.

Conservation agencies dispose of bachelor herds or floaters by trophy shooting to diminish male intraspecific competitors for breeding partners, which exacerbates fierce breeding contests during breeding seasons. This is obvious in gregarious animals like impalas, of which one male generally owns more than 30 females. When the number of men exceeds the number of females, there is unending fighting among males, which typically interferes with male eating and breeding periods.

Also, trophy hunting may function as an urgent control measure for human–wildlife conflicts that typically cause animal, human, and livestock fatalities; infrastructure damage; property and agricultural damage; and the spread of illnesses to either livestock or people. The prevalence and intensity of human–wildlife conflicts varied with species, nations, seasons, and the kinds of protected areas.

Human–wildlife conflicts occur primarily in the core protected areas. Human–carnivore conflict is one of the forms of human–wildlife conflict. There are approximately 226 carnivores in the world, almost all of whom are predators. In Africa, people have been interacting with carnivores for nearly four million years already. Consequently, crocodiles have preyed on humans and their forebears in Africa for roughly four million years.

Studies suggest that in Uganda, lions, leopards, and spotted hyenas caused approximately 393 fatalities between 1923 and 1994; crocodiles in Mozambique murdered over 300 people per year; and many more deaths remain unrecorded owing to communication problems. In the Kashmir area of Pakistan, leopards murdered 363 cattle between 2004 and 2007. Similarly, carnivores caused a financial loss of roughly USD 57,000 in northern Botswana.

In the United States, predatory predators killed more than 490 sheep and lambs, 83,000 goats, and 106 cattle, resulting in financial damage totaling two billion US dollars. Also, Ladan stated that between 1990 and 2004, lions had murdered roughly 563 people in Tanzania. Asian elephants have caused enormous agricultural destruction to farmers; for example, in India, animals damaged around one million hectares of crops and also demolished between 10,000 and 15,000 residences.

This brings the yearly agricultural damage caused by elephants in India to three million US dollars. Each year, elephants kill, injure, and destroy people's property throughout Africa and India. Every year, Indian elephants murder approximately 200 people in India. Elsewhere in Kenya, African elephants murdered 200 people between 2000 and 2007.

The district records in Tanzania show that African elephants killed roughly 40–50 individuals and wounded about 30–40 in 2009. Due to large economic losses, unending annoyances, and mortalities, several governments sanction disciplinary executions for responsible animals to lessen their adversative effects. In 2005, the Norwegian government destroyed the remaining wolf population because of significant predation consequences for sheep.

Likewise, in 2003, pastoralists poisoned all lions in Amboseli Reserve and speared roughly 27 lions in Nairobi National Park as punishment for murdering their cattle. Between 1920 and 1960, the Ugandan government murdered 106 leopards and 376 lions in an attempt to remove any troublesome animals. In addition, the Ugandan government rewarded carnivore killers.

Furthermore, rural people murdered 15 elephants in the Kilimanjaro area between 1996 and 1997 as a reprisal for agricultural damage, and 141 elephants were killed in Kenya's Amboseli environment between 1974 and 1990. According to studies, the trophy hunting alternative is cheaper, safer, more effective, and more profitable than most control units and local people. However, the technique is only applicable to male events because trophy hunting kills male animals exclusively.

In rare circumstances, the hunting of problem animals takes place during the hunting season when problems occur during the off-hunting season. The allocation of hunting quotas may be delayed due to bureaucracy. In certain nations, trophy hunting is timely and precise because the trophy hunting firms have pretty advanced resources to monitor the problem animals in various terrains, weather conditions, and times.

Even so, trophy hunting only restricts select species, not all species and individuals, since it primarily concentrates on animals with striking and big trophies. In other words, trophy hunting overlooks undesirable and unmarketable species and people. Also, trophy hunting frequently pushes severely endangered, juvenile, pregnant, and migrating species closer to the extinction notch. Also, it interferes with natural ecological processes such as animal migration and hydrological cycles. The existence of conservation corridors, water supplies, and migratory routes in the hunting fields interferes with animal natural migration.

The obstructed migration pathways enclose the migratory species in ecological islands, which in turn expose them to inbreeding and starvation owing to limited ecological supplies, especially in the dry seasons. The scenario became disastrous when hunting areas had improperly maintained tourist highways and trophy hunters drove off roads while seeking animals.

Most studies underestimate the cumulative consequences of off-road driving, notably on ecological processes, vegetation, and crawling species. The use of banned hunting equipment and tactics, such as guns capable of shooting more than one round of ammunition at once, typically damages the population of specific game species. Unethical trophy hunters deploy trapping snares that brutally kill the animals.

Unfortunately, trophy hunting excludes the world's most destructive pests, such as rats, *Quelea quelea*, and locusts, because they lack saleable trophies. As a result, governments exempt such species from hunting quotas. The problem animals managed by trophy hunting include lions, buffalo, leopards, elephants, and rhinoceroses. In the event of damaging effects, trophy hunting causes roughly 25% of species extinctions around the globe.

It caused the disappearance of the Dorcas gazelle (*Gazella dorcas*) and the Nubian bustard (*Neotis nuba*) from Sahelian Africa in the 1980s. Species extinctions frequently have multiple ecological, cultural, and socio-economic repercussions, particularly when trophy hunting includes a charismatic or keystone species.

There are plants, wild creatures, and human beings that fully depend on the overhunted species to thrive. For example, the local extinctions of elephants from various reserves led to the disappearance of numerous plant and animal species. Unselectively kill young, female, and critically endangered creatures. Moreover, the use of lead ammunition for trophy hunting directly impacts the trophy hunters and indirectly pollutes the land, water, and vegetation.

Trophy hunting eliminates people of a specified age, sex, and attributes from the population. It entails the extermination of physically fit and most morphologically pleasing animals. It wipes out most males responsible for protecting and leading the families to the safest feeding places. Furthermore, most breeding females choose their breeding partners based on their physical and genetic individualities.

Females may refrain from mating due to a lack of fittest males, triggering population collapse for the persecuted species. In certain cases, trophy hunting influences not only the population size and structure but also the sex ratio, fertility rate, and age structure by removing males and adult people from the population. There has been a fierce worldwide discussion on whether to cease or continue with tourist hunting.

Moreover, there is a misconception between legal trophy hunting and poaching owing to a lack of information among debaters. The Congressional Research Service (CRS) emphasizes that trophy hunting is the legal killing of an animal to obtain a certain part by following specified hunting regulations, while poaching is the capturing or killing of an animal for either a part of the entire animal with or without a valid hunting license.

Furthermore, CRS insists that a hunter with a valid trophy hunting license who hunts animals outside a specified area hunts animals with prohibited or restricted hunting gear, hunts off-season, hunts more animals than specified in the license, or hunts different types of animals or animals of different sexes or ages is normally described as a poacher.

Resident Hunting

It is a sort of sustainable wildlife tourism wherein residents and non-citizens are granted legal access to wildlife usage at a subsidized charge. It also grants legal rights to indigenous people to have legal access to wildlife utilization. The enabling legislation created different levels of fees for visitors and locals to hunt.

The state or regional governments create special hunting restrictions to allow citizens and traditional groups to shoot large game, birds, waterbirds, and invasive species. Locals receive access to natural resources through social equity, one of the fundamental foundations of sustainable development. It is completely subsistence hunting since it seeks to supply food and traditional remedies to the inhabitants. It is called "searching for a port."

The meat gained from resident hunting is normally not for sale, but it is consumed by the hunter and their family. This sort of hunting is for survival, not profit, and even the price of game species in resident hunting is generally subsidized to let inhabitants purchase the animals at inexpensive costs. For example, in Tanzania, African buffalo are sold for almost 45 USD in resident hunting, whereas in trophy hunting, the same animal is sold for more than 800 USD.

Resident hunting is more widespread in developing countries than in the developed world. Botswana, Namibia, and Tanzania are some of the nations with active resident hunting. Domestic hunting tourism occurs because some citizens travel from urban areas to hunting regions for many days in search of game species.

Despite the various viewpoints on history and trophy hunting practice, there is a link between trophy hunting and status. From an African viewpoint, trophy hunting is the killing of game animals for the acquisition of particular trophies of immense cultural worth. The African definition of trophy hunting focuses entirely on the kind, number, and quality of trophies rather than the hunting tactics and complexity of the weaponry used to kill the animals.

In that scenario, trophy hunting activity becomes significant to the satisfaction of traditional ceremonies among ethnic communities. Traditional cultures use the trophies they produce for coronations, weddings, traditional medicine, and decoration. It is an ancient cultural tradition in certain African nations with large animal resources. In Zimbabwe, Shona people, particularly the Manyika, see leopards as a royal game.

In the past, killing the leopard was forbidden for traditional healing and leadership. However, leopards were only slain when the famed traditional healers or traditional chiefs requested a skin. In other words, traditional hunters were required to obtain informal hunting permission from the traditional authorities, without which the leopard could not be killed. In essence, hunting for ancient rituals imposed comparatively less environmental damage than it does now since it was governed by customary regulations.

Hunters obeyed traditional regulations, which also regulated the hunting of particular species, but nowadays traditional hunting happens using random methods while inflicting tremendous environmental damage since it mainly targets rare or vulnerable species. However, most African nations have minimal supervision over traditional trophy hunting today.

Masai from Tanzania, for example, have murdered 2% of lions in the Serengeti habitat for traditional ceremonies. In the past, western civilizations rejected traditional hunting as trophy hunting, owing to the adoption of archaic hunting weapons and processes. According to this perspective, Western hunting culture recognized trophy hunting as a sign of Western civilization.

As a result, hunting regulations and procedures in the African hunting grounds still uphold the same concept by classifying game hunting into tourist and resident or local hunting, where tourist hunting is designed for highly paying tourists while resident hunting is for local people who hunt for meat only. Furthermore, hunters prefer wild to domesticated animals because it is a cage-free and organic meal.

It is a growing concern for inorganic or chemical agriculture around the world. People shun the intake of artificially generated food. Hunters prefer bush meat over domestically farmed meat because it is chemically unprocessed. The games lack artificial hormones, antibiotics, pesticides, and other problematic substances since they are reared in the wild.

Procedures for Resident Hunting

Universal hunting methods for resident hunters are non-existent since each hunting ground has a distinct hunting procedure reflecting its own culture, national legislation, accessible game species, and geographical surroundings. However, in other instances, resemblance may exist among the rules, such as subsidies on the price of game species, limits on hunted species, and locations.

Additionally, the commonalities in hunting regulations exist in all procedures, including familiarisation with hunting rules, obtaining a resident hunting license, preparation of hunting equipment, surveying the hunting areas, actual hunting, reflection of hunting exercise, and preparation for the next hunt.

Familiarisation with resident hunting laws is the first thing for all hunters to learn since everything about resident hunting is legally controlled. The hunter commits the offense against the state after breaching any of the sections or subsections of the hunting rules, with or without his awareness. In other words, engaging in resident hunting or any sort of hunting without the specified permission is as good as poaching.

It is also important that states or nations define and enforce safety and security restrictions for local hunters, the hunting destination, and the game species before, during, and after the hunting expedition. With that goal in mind, the rules establish all prerequisites for legally hunting the game species.

The resident hunting rule may set the age at which residents qualify as hunters. It is not a violation for a resident under the stipulated age to hunt or join hunting trips at a certain place. In most African hunting sites, including Tanzania, the hunter needs to be at least 18 years old. However, in the industrialized world, including the United States, a hunter must have at least 16 years of experience before obtaining a legal resident hunting license.

The resident is a perilous endeavor since it entails chasing dangerous wild creatures by utilizing harmful equipment like weapons. Allowing residents under the stipulated age to engage in any form of game hunting might be risky because of their inability to make quick and appropriate judgments in dangerous circumstances. Additionally, the hunting restrictions usually avoid contradicting the weapons and ammunition regulations that exclude people under the stipulated age from possessing and using firearms.

Also, the hunting rules allow the local hunter to understand precisely the sorts of game species that may be lawfully hunted since not all game species can be legally hunted. It is customary for hunting rules to prohibit some game species from hunting to respect their conservation status, sociocultural roles, and ecological responsibilities.

Most hunting rules restrict the hunting of critically endangered game species and national games. Furthermore, the restrictions discourage young and female hunters. Furthermore, the rules require hunters to have the correct hunting equipment for the right game species. For instance, the hunter should use the proper rifle and ammo to kill the specified game species.

In certain countries, like Tanzania, hunters are severely banned from using rifles that can kill huge creatures like African buffalo to shoot small animals like impala, and vice versa. Additionally, on certain hunting grounds, it is not legal to hunt wild animals with spears. In summary, familiarization with such regulations becomes vital for potential hunters since every hunting site has distinct hunting restrictions.

A comprehensive awareness of and familiarisation with hunting rules gives beneficial direction to hunters on the hunting processes, including getting a valid hunting license, preparing hunting equipment, scanning the area, a real hunting expedition, and reflection after hunting.

Like any other type of wildlife tourist activity, resident hunting necessitates that all hunters have a valid hunting license, issued by the authorized authority. A resident hunting license can be obtained at a private, communal, or public office, depending on the type of protected area or hunting ground. In such a scenario, the hunter needs to know where to purchase the hunting license.

The resident hunter requires a valid hunting license in the sense that the license should be issued by an authorized officer, the license should be used within its duration, the license should be used by the right person, the license should be used to kill the right animal, and the owner of the license should hunt at the right place as specified in the license.

Before participating in the resident hunting adventure, the hunter should arrange all hunting equipment in advance. The rifles and ammunition are not the hunting equipment the hunters should have, but everything from automobiles, food, water, field clothing, shoes, GPS, first aid kits, radios, a valid hunting license, and tents, among others.

Resident hunting may become complicated with massive, unexpected positive and negative occurrences. Maintaining and preparing the cars and other equipment in preparation helps the hunters minimize many unforeseen negative incidents during hunting journeys. Similarly, having the correct first aid equipment and expertise also becomes vital, as it may save the lives of hunters in the woods.

Also, inspecting the region before real hunting gives local hunters a hint as to what hunting gear and equipment they should bring along. Before hunting, hunters must be familiar with the hunting grounds, anticipating potential dangers and hunting success.

Hunting grounds can be scouted in a variety of ways, including meetings with experienced hunters and wildlife officials, field guides or manuals issued by hunting authorities, local hunting guides, and real-world surveys. Though having local hunting guides might be the most ideal alternative since the guides have been highly familiar with the hunting fields for years.

After knowing the hunting rules, acquiring a valid license, preparing the hunting equipment, and surveying the hunting region, the local hunter may then begin a real hunting expedition. The goal of this phase is to chase wild animals; this phase might endure for not less than a week. However, the hunting length may vary depending on the location.

That is to say, a resident hunting license gives the hunter at least a week to hunt permitted animals. The license's validity ends following a successful hunt or its expiration, whichever occurs first. It is fascinating to know what normally occurs if the resident's hunting license expires before a hunter acquires a kill. As indicated above, the hunter returns home empty-handed after daily searching for the licensed animal for the license period, which is at least 2 weeks.

Can the licensing officer or authority consider extending the license for a few additional days? Frankly speaking, the answer may be no, since most of the hunting grounds are well-informed about the abundance of hunted species in the hunting regions through frequent inspections, patrols, and reports from prior hunting seasons. As far as I am concerned, the hunter must use all of their efforts, energy, and resources to complete the kills before the license expires.

Hunters should use their abilities and expertise to study the actions and movements of the animals they are pursuing. Some of the techniques often used include stalking, ambush, and sitting in various blind spots, not where the animals drink water, waiting for the appropriate moment to strike and kill the animals.

After the hunting license expires, most licensing authorities will demand that the hunters return it. This is how the hunters submit their harvest to the authority. The returned license gives crucial information for protected areas to comprehend the animals killed, the site killed, and sometimes the days needed for the hunter to collect the kill.

Therapeutic Hunting

Therapeutic hunting is a form of consumptive wildlife tourism that is frequently employed to maintain and promote ecosystem health. It is a form of game hunting conducted to rescue or protect the environment under the umbrella of trophy, resident, or sport hunting.

In some countries, problem game animals are occasionally subjected to therapeutic hunting quotas to protect human life and property. In such an environment, frequent crop raiders or serial murderers are generally identified and slain to safeguard human life and property.

It is a three-in-one sort of hunting taking place during, before, or after the scheduled hunting season while protecting humans or other wildlife species from problem animals. The larger variety and numbers of predators surpassing the quantity and diversity of existing prey and the environmental carrying capacity of the protected areas may cause serious ecological risks through environmental degradation and killing of other wildlife species.

Moreover, the higher populations of wildlife species may cause problems for the communities residing near the protected areas by killing people and raiding crops and other human property. Previous studies assert that an increase in elephant populations in African landscapes frequently enhances human–elephant conflicts; the same applies to crocodiles and lions in East Africa.

In the United States, the excessive hunting of Burmese pythons is an effort by Florida State to lessen the aggression and serious harm the snake has inflicted on other animal populations since the 1980s. The python badly devastates lesser-sized creatures and produces ecological imbalances in the swamp. In brief, therapeutic hunting is a popular consumptive wildlife use among private wildlife ranches, farms, and private protected areas.

To reduce or entirely eradicate the unfavorable effects of problem species, protected areas may adopt game culling or cropping tactics. In this concept, the game hunters are assigned to search out and kill the wildlife creatures that are causing difficulties. The kind and amount of problem animals to be hunted may be publicly offered to hunting businesses or game hunters to compete.

Wildlife cropping as a wildlife management technique entails the harvesting or killing of wild animals to gain either game meat or control over overpopulated species and the acquisition of game meat. In all honesty, wildlife cropping and culling are ecologically sensitive tactics that demand careful implementation while avoiding unnecessary environmental, community, and social problems.

The game hunters may further sell the acquired quota to wildlife visitors in exchange for trophies or pleasure hunting. It is critical to note that therapeutic hunting is primarily and directly for ecological or ecosystem health but indirectly for recreational purposes. In any case, therapeutic hunting is like two sides of the same coin: the promotion of ecosystem health and wildlife tourist activities.

Fishing Tourism

Fishing tourism entails going to different fishing areas across the globe for the purpose of fishing. It is commonly referred to as recreational fishing or tourist fishing. It is a consumptive kind of wildlife tourism since it entails the complete off-take of different species of fish for consumptive or recreational objectives.

Tourist fishing is also a great way to discover aquatic ecosystems and recognize and eat many types of fish in marine, brackish, and fresh waterways. The Philippines is one of the most well-known tourist fishing destinations worldwide, providing a diversity of types of fish and fishing grounds.

Tourist fishing is an essential aspect of vacations, as many wildlife visitors spend their time on islands, riversides, lakesides, and the coast, as well as participating in fishing activities using fishing rods and other fishing instruments. Tourist fishing is one of the most recognized consumptive kinds of wildlife tourism, with significantly less debate surrounding its application.

Several causes led to the birth of fishing tourism, including a rise in tourist fishing operators such as fishing guides, boat owners, and resort owners, notably in the developing world. As an alternative kind of consumptive wildlife tourism, tourist fishing operators develop and deliver a particular fishing experience to wildlife visitors.

On tourist fishing trips, fishing operators and guides select specific sites and types of fish to attract and please a variety of nature visitors with varying fishing experiences, budgets, and interests. Moreover, the operators and guides enable wildlife visitors to get legal fishing permits and equipment for the effective chase of game fish.

Tourist fishing, despite being one of the oldest forms of consumptive wildlife tourism, lacks both a legal and institutional framework to govern its adoption and supervision, particularly in developing countries. In many fishing sites, tourist fishing is undeveloped, inadequately managed, and under-promoted. What is worse, even the contributions of tourist fishing to the economic, environmental, social, and cultural domains are still unclear.

Moreover, the lack of official data and inadequate knowledge of wildlife tourism players concerning tourist fishing are the key difficulties. Without a doubt, many sites with considerable water resources and well-developed fishing businesses are yet to embrace tourist fishing as part of wildlife tourism.

Sport Hunting

Sport hunting entails hunting a wild animal for leisure or killing it for fur. However, sport hunting is sometimes more complicated than it implies; it may further encompass the motives derived from resident and trophy hunting. Sport hunters may have a variety of pleasures or motives, ranging from the pursuit of a trophy to appreciation of nature and outdoor surroundings, task achievement, and honing their hunting skills and status.

In a sport hunting expedition, the major objective is to kill the animal for pleasure or recreational purposes. In this form of consumptive wildlife tourism, the hunt for a trophy becomes a secondary incentive. As indicated, the pursuit of wildlife trophies is optional. A sport hunter may devour or abandon the slain animal. On the other hand, it may sound like a violent or harmful form of enjoyment because it entails the killing of the animal without any purpose.

Apart from delighting in shooting the game species, equally sporty hunters appreciate nature and the outdoor mood of the protected terrain. It seems natural and satisfying for the trophy hunter to kill game species in an outdoor environment while enjoying the natural scenery, as it offers the actual feeling of nature and wildness. It also gives sport hunters the chance to vary their regular surroundings and alleviate daily living tensions.

Additionally, for different reasons, sports hunters may kill wild animals as a fulfillment of a professional or personal challenge or personal objective. A sport hunter may be interested in discovering and killing the toughest game species as merely a pursuit. Some sport hunters may want to understand how tough it is to shoot hazardous creatures like leopards or lions. It is out of such interest that wildlife visitors evolve into sport hunters, where they may kill the animals without taking anything from them.

Moreover, wildlife visitors may partake in sport hunting to strengthen their hunting abilities. Some novice hunters believe in the popular phrase "exercise is an excellent teacher, or exercise makes it perfect" and engage in numerous sport hunting safaris to shape their hunting abilities. On such safaris, wildlife visitors start learning to hunt peaceful animals before aggressive ones.

Furthermore, wildlife visitors view sport hunting as status. In certain communities, particular game animals—lions, buffalo, bears, and elephants, among others—are historically viewed as superior to others because of fairy tales and difficulties in hunting. For this reason, sport hunters recognize the successful killing of such animals as an emblem of brevity and status.

However, aside from the fundamental and secondary reasons or purposes of the sport hunters, the conservation authorities or protected areas also have specific contributions to sport hunting. As mentioned in other types of consumptive forms of wildlife tourism, conservation agencies may entertain sport hunting to dispose of habitual problem animals because of the constant damage to human properties and lives in peripheral communities, as well as manipulation of the wildlife population to avoid, among other ecological impacts, environmental degradation.

Protected regions must limit the number of regular crop raiders like elephants and velvet monkeys, as well as serial murderers like lions and leopards. Moreover, recreational hunting may be a useful tool for wildlife management to control and restrict overpopulated game species from exceeding ecological carrying capacity.

Furthermore, it supplies the conservation authorities with crucial information about age structures, sex structures, population numbers and distributions, and growth rates for hunting reports and monitoring. Crucial knowledge for wildlife management and protection. Sport hunting, like other forms of game hunting, creates various inconsistencies, including pain and suffering for animal species and species endangerment.

The damaged animals remain in terrible pain for a longer period before they die, owing to faulty shots, the use of unsuitable hunting equipment, or the use of traditional archery. Studies verified that only 20% of foxes hurt by hunters in the United States are shot again, while 80% stay unretrieved. The same applies to the three million wounded ducks that remained unretrieved. The abandoned or unretrieved injured people perish after a lengthy endurance of ceaseless aches, dehydration, malnutrition, and predation.

Moreover, game species are not the only victims of sport hunting; there are also human and non-game species. Every year, the National Shooting Sports Foundation of the United States claims thousands of injuries from sport hunting. The same applies to hunting areas in the developing world, where technological and regulatory infrastructure may restrict the timely data of such instances.

According to research, the cruelest kind of sport hunting is canned, caged, or restricted hunting, which usually happens in private protected landscapes. Canned hunting entails the slaughter of wild animals in enclosures or captivity. The proportions of the enclosure where the animal is hunted range from a few to thousands of meters.

This is the ***cruelest*** and most agonizing type of hunting because it does not give the hunted animal a chance to escape. It is a major wildlife tourism industry in the industrialized world, particularly in the United States. Ted Turner, the largest private protected environment in the United States, is one of the sites where thousands of wildlife visitors spend thousands of dollars to kill game animals.

Trophy Hunting Procedures

It is necessary to understand the protocols for game hunting; both trophy and sport hunting follow certain methods as defined in the hunting legislation. In some tourist hunting areas, tourist hunters may plan their hunting safari, while in other places, all hunters are forced to hunt via specialized hunting businesses.

Tourist Hunting is a recognized and approved hunting business entity that assists tourist hunting activities by providing housing amenities, hunting equipment, advice, obtaining permits, and transporting facilities to game hunters. The company may rent hunting block(s) from a public or private protected environment or own its own hunting block.

A hunting block is a geographical region where hunting activity occurs; it may be a whole or a small fraction of the protected areas. It is a thoroughly surveyed geographical region with enough quantity and diversity of hunted species leased to hunting firms with a defined term. Certain places, however, may grade their hunting blocks based on the diversity and availability of game species, ranging from one to five, with one being excellent and five being the worst.

The tourism firm with contracted hunting blocks obtains hunting quotas from relevant conservation agencies. It is the overall number, type, and number of hunted species for the upcoming hunting season. The authorities give, in advance, the list of animals to be hunted for each hunting season to each hunting business for each hunting block.

Each hunting firm consequently must promote and sell the earned quota to tourist hunters as a safari package. In addition to animal or game fees, the tourist hunting firm may include local transport, meals, conservation fees, concession fees, trophy preparation, lodging, equipment, advice, and beverages in each safari package.

The hunting rules may prescribe the type of equipment every hunting business should have before acquiring a tourist hunting company operating license. One might be the ownership of a certificate of firm registration, specialists including experienced hunters, and possession of necessary equipment such as all-wheel drive automobiles, freezers, birds, tents, and radio communications, among others.

Tourist hunting might be the costliest wildlife tourism activity in many wildlife regions since it entails direct and indirect expenses for wildlife tourists. When I was presenting about resident hunting, I stated that game fees for resident hunting in Tanzania are subsidized since it is about 10 times cheaper than tourist hunting.

That being stated, tourist hunting also entails professional hunting fees to lead wildlife visitors on hunting adventures. It is the professional hunting personnel who are aware of the hunting grounds, the searching of animals, the use of hunting equipment, and the hunting legislation on the hunting grounds.

Chapter 4
Impacts of Wildlife Tourism

Abstract This chapter addresses the advantages and disadvantages of wildlife tourism to various types of tourist stakeholders. Wildlife tourism involves a variety of important stakeholders, including local people, private tourism businesses, and the government, among others. The stakeholders benefit from wildlife tourism socially, economically, and environmentally. Apart from revenue generation and promotion of biodiversity conservation, wildlife tourism raises also several issues about a fair distribution of benefits among its players, resource management in the protected areas, and sustainability. In certain instances, balancing between ecological integrity and the rise of wildlife tourism becomes challenging owing to the incapability of sticking with the moral norms and principles of sustainability. Wildlife tourism is an essential economic sector, sustaining the local economy and encouraging conservation initiatives in the host countries. The success of wildlife tourism hinges on its ability to meet the interests of all stakeholders without damaging the biological integrity of protected areas. Involvement of all wildlife tourism stakeholders from local, regional, national, and international perspectives may enhance the execution and advantageous effects of wildlife tourism. An effectively implemented wildlife tourism delivers various benefits to its stakeholders, including the potential for employment, the creation of social infrastructure and superstructures, and the conservation of protected areas and keystone species. However, unregulated wildlife tourism may become socioeconomically, culturally, and ecologically compromising, as it could lead to poaching incidents, commercialization of culture, and increased poverty by turning residents of local communities into beggars. Conservation agencies, tourism enterprises, and government entities must collaborate by working together to minimize the detrimental effects of wildlife tourism. Maximization of societal, cultural, and financial implications of wildlife tourism might increase the viability of wildlife tourism. The effective and transparent management of the funds accrued from wildlife tourism could motivate and maximize community involvement, facilitate the implementation of conservation measures in the protected areas, and facilitate law enforcement and research for preserving endangered species.

A. A. Mamboleo, *Sustainable Wildlife Tourism*,
https://doi.org/10.1007/978-3-031-97795-4_4

Keywords Wildlife tourism · Sustainable tourism · Stakeholder engagement · Biodiversity conservation · Economic benefits · Human-wildlife conflict

Positive Impacts of Wildlife Tourism

This chapter introduces the tangible and intangible benefits of wildlife tourism from the institutional to the household levels. It further discusses the social, economic, cultural, and environmental impacts of wildlife tourism in developing and developed countries.

Wildlife tourism is a form of tourism that includes hunting and observing wild animals, plants, and their related natural ecosystems. Wildlife tourism is a significant aspect of the tourist businesses in several African nations, such as Tanzania, Kenya, Uganda, Botswana, and South Africa, to name a few. Since our book focuses on sustainable wildlife tourism, it is crucial that we also look at the advantages of wildlife tourism to wildlife stakeholders, including local people.

There are many stakeholders in wildlife tourism; if we had to conduct a stakeholder analysis, it would take us thousands of pages to detail each of the key wildlife tourism stakeholders. Let us do it by defining who is a stakeholder: a person or institution that influences or is impacted by wildlife tourism. Different layers of stakeholders may exist, with some of the most common being geographical and affiliation-based. Geographically, stakeholders in wildlife tourism may be local, national, regional, and international.

Wildlife that is transnational or transboundary, such as international migratory birds and mammals, may cross local, regional, and international borders, influencing multiple nations and continents. Wildlife tourism is also the product of local, regional, and international travelers.

On a local and geographical scale, wildlife tourism impacts local people in terms of conservation costs and benefits. That is to say, stakeholders are highly essential in wildlife tourism since their involvement in wildlife tourism may produce good and bad consequences for wildlife and wildlife tourism.

By definition, we have quickly clarified the word stakeholder and restricted ourselves to select a few instances. As I said earlier, every stakeholder has beneficial and bad effects on wildlife tourism and wildlife species. As far as sustainable wildlife tourism is concerned, wildlife destination management needs to conform to the principles of sustainability to progress wildlife tourism to sustainable wildlife tourism. The managers must restrict the advantages of wildlife stakeholders within natural resources since the acquisition of the benefits must respect the capacity of the environment.

In a nutshell, the kind and amount of advantages each stakeholder receives from sustainable wildlife tourism may drastically vary. For example, the benefits of hotels may differ from the benefits a protected area receives from wildlife tourism. That is to say, the interests of many hotels are commercial-focused, whereas protected areas are ecological- and cultural-orientated.

Apart from environmental deterioration, the principles of sustainable development describe local people as a disenfranchised population with little or no access to natural resources. Despite being closer to protected regions, they have heavily regulated access to natural resources. In that respect, the explanations of the advantages of wildlife tourism may be classified by local people and other stakeholders.

Socioeconomic Benefits

This section illustrates the advantages that various stakeholders may receive from wildlife tourism. We start with other stakeholders so that we may have the wide knowledge that stakeholders normally receive from wildlife tourism. Income creation is one of the most critical.

Wildlife tourism companies and housing facilities earn trillions of dollars a year. It is visible in renowned wildlife tourism destinations, especially in developing nations with intact and distinctive animal resources. Such investments generate employment for local and international tourism specialists.

Employment Opportunities

Wildlife tourism provides direct and indirect employment opportunities for local and foreign workers. Also, employment occurs both officially and informally, since some individuals are hired via formal tourist organizations and others through informal tourism institutions. However, there have been fewer considerations for local workers in the tourism business, which was originally caused by poor or unconnected academic backgrounds and experience.

Currently, most employees are encouraged to safeguard the local employment market by employing local tourist professionals to fill relevant roles before recruiting foreign ones, as most workers in developing nations choose to hire local experts only when filling low or informal working positions. The creation of job possibilities is one of the most significant components of wildlife tourism, but evaluating whether the chances are fair, legal, and ethical is vital.

Because there was a period when tourism was suspected of being one of the biggest sources of child employment, more than 200,000 children were unlawfully working in the tourist business. Having such unethical and unconducive working conditions may neutralize all the employment advantages provided by wildlife tourism. Similarly, an uneven distribution of work prospects across different socioeconomic groups and education levels may fail to justify the contribution of wildlife tourism to employment chances.

That brings us to another question: whether inconsistencies in job prospects may degrade the sustainable image of wildlife tourism or not. There is no straight rebuttal to the statement since sustainable wildlife tourism is the result of three main pillars or principles of sustainability, among them social equality. When job opportunities challenge social equity and respect, wildlife tourism may be unable to thrive.

Similarly, inadequate assistance or empowerment for the informal sector in establishing local job possibilities is critical. When informal and private sectors may provide various job chances when supplied with the opportunity to function properly in the wildlife tourism business.

The informal sector may be encouraged to create certain items or deliver services to assist in the running of the wildlife tourism system. For example, locals may be routinely educated to give particular hotels commodities such as fruits, vegetables, and decorations. Furthermore, interpretive services, such as tour guides in neighboring wildlife areas, are required.

Business Investments

However, the benefits to other stakeholders are unclear because, in most cases, they tend to benefit more international investors than local investors. This has been the case in developing countries, where capital and investment policies continue to favor foreign investments. From a sustainability standpoint, it is crucial for investment policies to equally consider strengthening local investors as a method of providing a level of competition with international investors.

In certain nations, defined investment rules offer direction for local employment and wildlife-related investments. For instance, the policies may demand that the firm be ecologically and socio-economically responsible. In that case, every investment must contribute a specific proportion of accrued earnings to communities and conservation agencies. This helped reduce the detrimental repercussions of enclave tourism, which was the primary aspect of many wildlife tourist investments in protected regions.

With enclave tourism, tourist investments have minimal or no contact with other tourism stakeholders, including residents. In that sense, the isolation of certain wildlife-related economic ventures from other stakeholders overlooks the consumption of local goods and services. That is to say, enclave tourism undermines the socio-economic development of local communities and the country at large, which is opposed to the ideals of sustainability.

In various places, there have been cases where overseas visitors regularly pay for housing, local and international transit, meals, insurance, and other services abroad. More unfortunately, after arriving at the location, travelers are supplied with little opportunity to contact local populations to squander their money on souvenirs and other locally manufactured items. Moreover, such separation makes the wildlife-associated commercial businesses unfair and insensitive to the local economy, and it is utterly against the principles of sustainable development.

According to the principles, wildlife visitors and operators must encourage and empower local people's efforts toward socio-economic development while prohibiting any type of environmental degradation. The lack of both social and financial exchanges with overseas visitors hinders local aspirations toward economic freedom and self-reliance. However, several governments and wildlife tour companies have genuinely learned from previous blunders and adjusted their rules to address such failures.

Because of such lessons and coordinated actions, some wildlife operators have begun investing in village lands to keep their businesses closer to communities. In the course of their activities, the investors engage closely with locals by enabling the communities to establish ecologically and socio-economically appropriate enterprises that may attract wildlife visitors.

Such firms acquire ethically locally available items from the locals at the fair area and arrange sometimes for visitors to engage with local people. Furthermore, the firms provide monthly training to villagers on goods and service management to meet the requirements. The company may even hire local tour guides for animal activities inside and outside of the protected zones.

Moreover, trophy hunting causes less economic leakage than other forms of tourism; around 92% of trophy hunting firms are situated in the nations where the hunting takes place (Lindsey, 2015a and b). In such cases, most of the income gained from trophy hunting firms stays in the hunting locations. However, this is not the case for nations all over the world, as several African hunting grounds fail to regulate the large tourist leakage.

In addition, high-spending trophy hunters may lead to increased economic activity in the community's neighboring hunting areas (Van der Meer & Dullemont, 2021). As a result, trophy-hunting activities improve both direct and indirect job prospects in communities (Hasler, 2000; Lindsey, 2015a and b). Yet, the allocation of trophy hunting earnings may differ across hunting sites.

The trophy hunting income sharing between the governments and communities spans between 0 and 100%, and the communities may receive between 0 and 100% depending on the country's policy. Tanzania, for example, invests 25% of trophy hunting profits to promote community development programs.

Revenues Generation

Although the primary goal of whether wildlife resources are protected publicly, privately, or communally, wildlife tourism is also essential. The protected regions earn money from wildlife tourism activities, which is critical for funding conservation initiatives.

I remember highlighting in the previous pages that finance was one of the most important things to consider when establishing the protected areas. Without an adequate allocation of funding, the administration of protected areas may be challenging. You might think that law enforcement actions and the creation and maintenance of tourism infrastructure and amenities in protected regions are impossible without reliable sources and an adequate amount of money.

Protected areas earn money through varying admission fees, concessions, and the sale of animals to trophy hunters. Wildlife tourism is a complicated industry requiring numerous activities and authorities, such as the registration and certification of tour firms, the certification of tour professionals, and the rental of wildlife regions for wildlife tourism businesses.

All these efforts are expensive, as they entail the payment of a specific amount of money to the government and certain protected regions as well. However, the adequacy of the accumulated money for reinvestment in conservation may change depending on the protected areas' rules and ownership.

In publicly held protected areas, it is specifically stated how much wildlife tourist income should be reinvested in conservation. For example, Tanzania reinvests 75% of the money from tourist hunting back into wildlife conservation, and the balance goes to the people around wildlife regions. This may operate as opposed to privately held protected areas, which greatly rely on money from wildlife resources to maintain everything in the protected area.

In certain situations, the profits from animal resources may be less than the protected area's real yearly budget. For publicly held protected areas, either the government may alleviate the budget or the contributions from the donors, such as WWF, USAID, and FZS, among others, may top up the budget. Other protected areas may cover the budget for those governed by one organization.

Medication Values

Wildlife tourism is a source of medicine since it delivers medicinal benefits to local and foreign travelers. Visitations to protected sites may give direct and indirect treatments to visitors inside the protected areas. You may have heard about thermal baths and waterfall viewing.

The protected places with thermal resources and waterfalls give spiritual and psychological healing to certain visitors. I recall my journey to Lake Manyara National Park as a student. I met the park official, who told me that one of the hot springs in the park draws hundreds of visitors with Asian identities who take baths as their spiritual therapy.

Thermal tourism is a form of medical tourism that involves using subsurface hot or cold water to treat or prevent mental, joint, and inflammatory illnesses. Thermal medicine involves mud baths, natural spas, inhalation therapies, and even drinking.

It is an old tradition that natural hot water, rich in minerals and gas, is full of therapeutic and medicinal benefits. The geopolitical settings of various nations, including the protected regions bisected by rift valley lakes in East Africa, such as Tanzania and Kenya, are among the most favorable for their protected areas for thermal tourism.

Turkey, with more than 1600 hot springs, ranks first in Europe and seventh on the globe as a potential thermal tourist destination. Similarly, nations like New Zealand, Italy, Spain, Germany, and Iceland have been noted for their long-lasting expertise in thermal tourism.

The aboriginal Indians of America and the indigenous people of Australia see hot springs as sacred locations. In some circumstances, thermal resources in protected places may be employed as visual attractions, spiritual attractions, or health and wellness spas because of their healing capabilities.

The rift valley lakes of East Africa, such as Lake Natron, Bogoria, Naivasha, and Manyara, most of which are situated in protected areas, have the potential to become Africa's most renowned thermal wildlife tourist attractions. As previously stated, the Lake Manyara National Area sought to accommodate Asian tourists visiting for thermal baths over Christmas.

Additional lakes, such as Lake Ngozi in Tanzania, which is part of the Poroto Forest Reserve, are another key draw for thermal tourism. More significantly, many waterfalls in the protected regions may also become key tourist, recreational, and medical attractions. Watching every water drop fall from the top to the bottom brings psychological comfort to certain individuals.

In their research, Jo et al. (2023), the findings verified that a waterfall picture promotes sympathetic nerve activity, a happy mood, and relaxation. Therefore, it was concluded that waterfalls may create physiological waking and psychological soothing effects since, for many years, water has been used for therapy of mental and psychological issues.

Therefore, gaining psychological healing via hydropower. It is a crucial aspect of tourism, especially in our period when many people are burdened with everyday problems of existence. It is unofficially integrated as a tour package in many wildlife regions, notably in underdeveloped nations where stress is not seen as a concern. Despite the endowment of various waterfalls in the wildlife-protected areas, many wildlife destinations barely view waterfalls as a component of medical tourism and so hardly spend money on selling them as an independent package.

Benefits to Local People

Local people living near wildlife tourist hotspots are the major victims of wildlife conservation. The conservation costs impoverish their lives since residents lose their property and lives via predation, crop raiding, and many other indirect repercussions. In some cases, the benefits shared by wildlife tourists should always outweigh the costs of suffering.

Through the development of social infrastructure, African governments have sought to share the advantages of wildlife tourism with the surrounding people. Other governments encourage local people to manage wildlife on communal grounds via community-orientated natural resources initiatives. Such decentralization permits locals to lawfully manage and exploit wildlife resources for socioeconomic development.

The transition of the natural resources management paradigm from centralized or conventional to community-based management led to the engagement of the then-disadvantaged local communities in community-based tourism. Through cooperation, the animal resources on common land have become significant attractions for wildlife tourism.

Surprisingly, certain community wildlife areas offer vastly more appealing and distinctive resources than privately and publicly held protected areas. The then-Wami-Mbiki Wildlife Management Areas had a larger surface area than other protected areas in Tanzania.

Even resource-wise, the region had more biological assets than certain protected places, notably Saanane Island National Park. The community-owned protected area was larger than the combined surface area of Kilimanjaro National Park, Lake Manyara National Park, and Arusha National Park. As a result, it was elevated to game reserve status.

Community-based organizations can manage animals on their territory through wildlife tourism. The community spaces provide direct and indirect employment opportunities for locals and other specialists. Furthermore, the proceeds from wildlife tourism are used directly for local development.

The people actively plan for resource management and protection, with little or no outside influence from other wildlife conservation groups. As a result, local people develop ownership of natural resources and may sometimes perceive animals as their own. However, most community wildlife programs are still under investigation because they fail to capture their direct contribution to household levels.

Locals enjoy institutional advantages in many community-based wildlife schemes, such as the building of schools, dispensaries, village offices, and police stations, although some locals barely see such projects as part of community-based wildlife tourism. Some local people say governments have to create social infrastructure, and thus they want the money from community-based wildlife tourism to be used for the development of household projects.

Taxation and Licensing

Governments also typically accumulate money through the issuing of permits, licenses, concessions, and taxes. It produces more earnings per visitor than any other kind of tourism in sub-Saharan Africa (Lindsey, 2015a and b). Lindsey and colleagues suggest that trophy hunting produces 30 times more money than photography and ecotourism in Zimbabwe and Tanzania.

Semcer (2019) reported that trophy hunting in Africa generates monetary earnings ranging from USD 190 million to USD 326.5 million per year. According to Semcer, even at the lower end of the expected income production, profits from trophy hunting are still about one-third larger than the USD 142 million earned as admission charges from ecotourism in 14 African nations.

Governments frequently reinvest a percentage of the funds generated in sustainable wildlife protection. A successful and timely reinvestment makes such income demonstrate the "biodiversity umbrella effect" as conservation authorities employ them for the conservation of non-hunted species as well.

Reinvestment of earnings in wildlife conservation typically defines the efficiency and validity of trophy hunting's contribution to conservation. The more money invested, the more significant trophy hunting becomes in animal conservation.

Environmental Impacts of Wildlife Tourism

This section introduces the link between wildlife tourism and the natural environment. The complex link between tourism and the environment causes tourism activities to have both negative and positive environmental effects. Wildlife tourism as an environmentally constructive tool, as well as sustainable wildlife tourism, may support environmental conservation activities directly and indirectly through the provision of conservation funds, the promotion of civilization, species conservation, the conservation of scenic features, and local and international donations.

Conservation of Wildlife Species

Before talking about the benefits of wildlife tourism for the conservation and control of protected areas, it is important to briefly explain some key terms in the field of nature conservation. It is common to come across difficult words like rare, indigenous, keystone, endangered, dead, and threatened species when either talking to environmentalists or reading conservation books.

Similarly, when watching nature-based movies or shows, it is common to come across words like "The black rhinoceros is among the nearly extinct species in Africa." Rare species reflect the groups of wildlife species that are usually low in number. In that context, human proportions have either little or nothing to do with the size of such species, as their numbers are naturally small.

Though their smallness carries the risk of vanishing shortly, special conservation efforts are needed to maintain the size and survival of their populations. However, people and non-human causes can still pose a risk to the disappearance of rare species.

Rarity causes a shortage of those species, making them unique and exceptional wildlife tourist resources. Pangolins are one of the rare species with several beautiful traits and socio-cultural features. In some countries, such as Tanzania, people associate the appearance of a pangolin with luckiness.

Rare, unique, and threatened wildlife species will be handled and conserved properly. Wildlife species, a key tourist draw, will necessitate that tourism partners increase their conservation efforts to reverse their declining trends. Tourism partners, such as tour companies and ecological hotel owners, actively participate in protecting threatened species.

Wildlife tourism supports the protection of core, base, and warning species. Gorgeous species such as elephants, rhinos, giraffes, baobabs, and blackwood will always be preserved for tourists. It also improves the conservation and control of attractive landscape features such as beaches, mountains, lakes, and rivers.

Revenue produced from tourist hunts and picture tourism goes to the Ministry of Tourism and Natural Resources in Tanzania. This kind of tourism produces the bulk of the income in direct profits from Tanzania's Wildlife Division from reserved land through a variety of fees. However, most of these revenues are returned to the

Central Treasury, and only 25% is given to the Tanzania Wildlife Protection Fund, indicating that tourist hunting generates more revenue than other tourism.

Protecting all the world's threatened species, including the most hunted species on hunting grounds, costs around 4 billion US dollars a year (Cressey, 2012). The costs seem very large, but in terms of government budgets, species abundance, tourist activities, financing, and the country's surface area, they are quite reasonable.

Cressey cautions that annual expenditures are a fraction of the value of nature's ecosystem services. However, there is a notable variation in the extent of revenue reinvestment in conservation across countries, with some governments or agencies investing more than others. Tanzania, for example, reinvests 25% of hunting revenues into wildlife protection through the Tanzania Wildlife Protection Fund.

In some countries, the mushrooming of community-based natural resources management (CBNRM) schemes and private reserves has overshadowed competing land uses, such as livestock keeping, urbanization, and farming. In other words, proprietors transform their property into wildlife reserves that enhance the landscape for species conservation and opportunities for tourism-based benefits.

Furthermore, it engages proprietors in wildlife-based conservation and restoration initiatives, such as anti-poaching. In Namibia, trophy hunting resulted in the establishment of communal conservancies encompassing nearly 14% (12 million hectares) of the country's landmass.

WWF reported that 19 established WMAs added 3% to the total surface area reserved for wildlife in Tanzania, which now totals 31%; however, when the other 19 proposed WMAs are gazetted, WMAs will increase the total land area under community conservation to over 10% (about 300,000 km^2) of the country's surface area.

In 1980, Zimbabwe dedicated only 12% of her land to wildlife administration, all within government-protected areas.

After the inception of the CAMPFIRE program, the reserved land in Zimbabwe comprised 33% up until 2000. In other words, the program added 21% to the entire reserve land in Zimbabwe; in terms of surface area, the CAMPFIRE program outweighs more than twice the total reserved area covered by the government reserves.

Therefore, private hunting operations in Africa manage about 1.4 million km^2 of land, representing 22% more land than is protected by national parks. In other words, the size of private hunting grounds in Africa, which is also marginalized land, is more than three times the size of the United States National Park System, roughly two times the size of the United States National Wilderness Preservation System, and more than twice the size of the United States National Wildlife Refuge System.

Local and International Donations of Resources

Wildlife tourism attracts direct and indirect financial donations for wildlife conservation as well as the socio-economic development of local communities. Individual visitors and organizations offer various types of resources for species management or protected areas. International and local organizations, as well as individuals, donate funds and working instruments to the government and communities for the in-situ and ex-situ conservation of wildlife resources.

It is evident to many responsible wildlife travelers to learn and experience the reality of the conservation challenges that conservation organizations and communities nearby the conservation areas undergo during conservation. It is based on this fact that wildlife travelers decide to contribute responsibly to support the ongoing conservation initiatives through either conservation organizations or individual efforts.

The contributions from wildlife visitors occur as financial or non-financial entities. Conservation agencies, particularly in Africa, may receive technical assistance from various sources as a consequence of wildlife tourism activities. Technological and logistical assistance to fight back some of the ranking conservation challenges, including human-wildlife conflicts and extraterrestrial invasive species, are essential to conservation costs as they are financially and technically costly.

The involvement of responsible wildlife visitors and wildlife tour operators encourages conservation agencies, as some governments have insufficient budgets to finance all conservation initiatives flawlessly. The large surface area of protected areas' networks and the feeble economies of some countries may contribute to such inefficiencies. Contributions from wildlife visitors, organizations, and operators may target the entire ecosystem, a protected area, or certain species.

Wildlife travelers or operators may be concerned with the conservation challenges of a particular wildlife species and opt to support the ongoing conservation efforts to save the species from extinction. In the previous sections, I mildly described several species terminologies, including threatened and endangered species.

With this kind of donation, wildlife tourism may persuade the stakeholders to rescue the threatened species through donations of the required resources. Also, the conservation of an entire ecosystem or protected area has become the interest of some wildlife tourists and operators.

In that case, responsible visitors may voluntarily donate their resources to support the conservation of all species and the environment in the particular protected area. Representativeness and long-term support of protected areas to local and international communities persuade the wildlife tourism stakeholders to donate resources for supporting conservation efforts.

Wildlife Tourism Research and Monitoring

In the previous chapter, the author cited the statement from Prof. Songorwa emphasizing tourism as a fully loaded gun in the hand of a mad person, though this was the way the professor found it easy to make his students understand better the worst adverse impacts of unmanaged tourism operations on protected areas, communities, and the economy.

I don't believe Prof. Songorwa is the only person who understood and endeavored to convey the worst impacts that tourism can have in such a different manner. Frankly speaking, the statement from the professor convinces some of us that tourism operations are destructive if not controlled and monitored properly and timely.

It is essential to clarify that tourism operations are destructive, as they require the preparedness and proactiveness of relevant destination and protected area administrators to anticipate and control most of the impacts before they become real. In that case, timing and proactiveness are keys to regulating the impacts. However, comprehension and anticipation of the prospective impacts of wildlife tourism are difficult without the availability of correct information.

One of the primary sources of such information is scientific research to comprehend the consequences of wildlife operations. For example, the increased number of wildlife tourists normally challenges conservation authorities. This motivates intensified socio-economic and ecological research to facilitate routine monitoring of wildlife activities.

Wildlife tourism necessitates that wildlife stakeholders, including conservation agencies and destination management organizations, conduct various research to understand various behaviors and potential consequences of wildlife tourism. As a method of avoiding such consequences, routine research aids in comprehending the problems and suggesting potential mitigation measures. With the right information about the consequences and also the correct scientific mitigation measures.

Reduction of Poaching Incidents

Wildlife visitors may directly or indirectly help to reduce poaching incidents in wildlife regions. It is feasible for wildlife tourists to directly reduce the poaching pressure as they make wildlife attractions active with tourism activities, therefore denying the opportunities for poaching incidents. That is to say, the busiest wildlife areas will scarcely be visited by poachers.

Tourist activities make the destinations congested and frighten wildlife tourists away. More importantly, at some wildlife attractions, various night wildlife tourism activities, including night game drives, may assist in frightening away night poaching operations. Indirectly responsible wildlife travelers encourage wildlife tour operators, conservation agencies, and other wildlife tourism stakeholders to adhere to sustainability principles.

Because responsible wildlife travelers may hardly visit protected areas with frequent and violent wildlife crimes, as a result, stakeholders become environmentally responsible by similarly undertaking routine anti-poaching operations to save wildlife resources. As a part of anti-poaching operations, conservation agencies, and wildlife tour operators may also educate communities and travelers about the impacts of wildlife crimes.

Furthermore, some wildlife tour operators have special foundations in place to directly or indirectly support anti-poaching operations in private and public protected areas. The operator may actively partake in the anti-poaching operations. As wildlife tour operating companies own and manage wildlife areas with sufficient wildlife resources, where wildlife is used as dispersal and migratory areas, the wildlife tour companies conduct regular operations to safeguard the wildlife resources.

The wildlife operators employ qualified personnel and deploy state-of-the-art anti-poaching equipment to fight back against wildlife crimes. As part of their environmental engagement policies, some wildlife tour operators indirectly support conservation agencies' anti-poaching activities by donating financial and non-financial resources to relevant organizations.

Such donations empower public and private environmental agencies to regulate wildlife exploitation. Anti-poaching is one of the primary functions of many protected areas, but several conservation agencies in developing countries have limited budgets and equipment to adequately conduct anti-poaching operations. Additionally, implementing a community-based wildlife tourism policy may reduce wildlife exploitation.

Permitting local people to wisely utilize the wildlife resources on their land through wildlife tourism may enhance their sensibilities toward wildlife ownership and similarly alter their negative perceptions of wildlife. If the benefits of wildlife tourism offset the adverse impacts, it may encourage the locals to protect wildlife resources against wildlife crimes, both directly and indirectly.

Villages with prominent wildlife resources in their territories may directly participate in anti-poaching operations to safeguard the wildlife. Some governments permit and empower villagers to manage wildlife resources within their territories. Wildlife tourism, as one of the authorized activities in their localities, encourages villagers to continue safeguarding their resources for income generation.

The village may establish a community-based organization to manage daily conservation and tourism activities. As a result, the organizations may officially employ village game trackers and wildlife experts for anti-poaching operations and tour guidance. Wildlife tourism encourages local people, as exemplary Samaritans, to report any wildlife crime occurring near or within the protected areas to relevant authorities.

Communities living closer to protected areas are sometimes aware of wildlife offenses and offenders residing in their communities. When wildlife tourism is appropriately motivated by the construction of social infrastructure and household economic capital, local people may be willing to collaborate with law enforcement organs in anti-poaching campaigns.

Promotion of Civilization

Wildlife tourism entails bringing people from diverse cultural backgrounds together; it provides wildlife travelers with opportunities to observe and learn about different cultural experiences. That means that through wildlife tourism, people learn about a variety of life experiences, including dining, worshipping, and attire methods, among others.

In that scenario, wildlife tourism becomes a platform for learning and exchanging cultural experiences. Wildlife tourism enables exchanging various cultural experiences by imparting and acquiring beneficial habits between wildlife tourists and the host communities.

All these are crucial for advancing the ethos of interacting parties. Learning constructive behaviors on the other side may enhance the culture of the host community and vacationers. The exchange of valuable cultural experiences among visitors in different cultural contexts is essential for socioeconomic development and environmental conservation.

Wildlife travelers learn about their lifestyles from other tourists or natives. Tourism may alter people's acceptable lifestyle by restricting the purchase of products from endangered species, dietary practices, and their perceptions of biodiversity. Wildlife tourism may increase the visibility of certain cultures or ethnic groups nationally, regionally, and globally.

The cultural attractiveness extends to its locality, enticing many visitors. In this context, any culture associated with fauna can readily become conspicuous and alluring. It is commonplace for wildlife travelers to comprehend the way wildlife influences or is influenced by particular cultures.

In the East African region, the Masai culture is well known globally for several things, including its association with the history of wildlife and protected areas in the region. In the past, certain protected areas or wildlife habitats were Masai's ancestral residences, and the way the ethnic group interacts with species is exceptional. Without mentioning the Masai, it is impossible to tell the annals of wildlife conservation or tourism in East Africa.

In the past, the Masai coexisted with docile and hostile wildlife species without causing any damage to either party. However, some protected areas still evidence such coexistence. Take a look at the Ngorongoro Conservation Area in Tanzania, where the Masai still coexist with untamed animals. Such coexistence makes Masai culture unique as it raises many doubtful concerns about the potential consequences for their livelihoods, livestock, and wildlife.

Such inquiries inherently increase wildlife travelers' curiosity by requesting more information about Masai culture. Furthermore, wildlife travelers may share any information obtained about the uniqueness and beauty of Masai culture with friends and relatives. The fascination and interest of wildlife visitors may similarly create cultural awareness of the environmental and socio-economic significance of the culture among the Masai themselves, governments, and other tourism stakeholders.

With Masai as an example, certain aspects' unique behaviors have become both an attraction for wildlife travelers and a lesson. As an attraction, the culture may draw thousands of tourists from various walks of life to venerate it, as it is part of Masai culture. It becomes a lesson because it is fascinating to see Masai coexisting with fauna with little or no damage at all.

It serves as a lesson to conservation agencies on how they can reduce incidents of human-wildlife conflicts with only indigenous knowledge. But more importantly, it helps wildlife enthusiasts comprehend the importance and possibility of interacting with nature without any damage. Moreover, wildlife tourism promotes social transactions, where various people may interact physically and emotionally.

This provides an opportunity for wildlife travelers and host communities to understand and learn more from each other. Through their interactions, wildlife travelers may clarify some cultural misconceptions. A thorough understanding of each party promotes harmony, tolerance, and socio-economic development.

Adverse Impacts of Wildlife Tourism

It indicates all the undesirable effects that tourism activities have on the environment. There is a complex relationship between tourism and the environment, and many activities can have adverse environmental effects. The complex relationship between tourism and the environment causes tourism activities to have both negative and positive environmental impacts.

The development of tourism infrastructure and facilities such as highways, airports, resorts, hotels, restaurants, and golf courses can have negative impacts on the destination. However, tourism has the potential to create beneficial effects on environmental protection and conservation. It is a method to raise awareness of environmental values, and it can serve as an instrument to finance the preservation of natural areas and increase their economic importance.

Wildlife tourism can cause significant disturbances to animals in their natural habitats. The increasing interest in traveling to developing countries has caused a surge in resort and hotel construction, particularly on rainforest and mangrove forest lands. Wildlife observation can drive animals away, disrupt their grazing and reproductive sites, or acclimate them to human presence.

Infamous wildlife tourist destinations, including Botswana, Namibia, and Tanzania, wildlife observers drive keystone and endangered species off their reserves, increasing the risk of inbreeding and further intensifying the interactions between humans and wildlife. In brief, wildlife tourism has socio-economic and environmental impacts.

Destructive Impacts on Environment

Due to the nature of wildlife tourism, which comprises interactions between humans and natural vegetation and humans and wild animals, disease transmission is undoubtedly conceivable. The question may be, where are the primary origins of the diseases being transmitted to humans or wildlife?

Transmission of diseases in wildlife tourism can be both from wildlife and from wildlife tourists, who can transmit disease-causing pathogens. Wildlife has been known as a disease reservoir, but in many cases, it has co-evolved with some diseases and created a balance. New diseases introduced to wildlife can be a catastrophic problem.

In this context, wildlife tourists can be the primary transmitters of diseases among wildlife. Wildlife tourism involving the collection of animals in the same areas is likely to increase the transmission of diseases between wildlife tourists and wildlife.

As it was noted earlier, wildlife remains the primary reservoir of various disease-causing pathogens, accumulating and keeping several animals together in private protected areas, wildlife estates, wildlife farms, or zoological gardens. That concept is like accumulating many hazardous pathogens at the same location.

Since each maintained species has a specific disease-causing pathogen, wildlife visitors can acquire most of the pathogens from the wildlife. The same wildlife visitors may transmit pathogens to wildlife through heightened interactions. Moreover, as we analyze the impacts of wildlife tourism, we have noted that some wildlife tourist activities, like wildlife viewing and hunting, are inherently destructive; they exacerbate natural habitats and scare wild animals away from their natural areas.

As a result, the pursued animals travel across various protected and unprotected land uses, including the human landscapes; it is where the exchange of harmful disease-causing pathogens, such as anthrax, Ebola, and rabies, among others, between humans, wildlife, and livestock, may occur.

It is a common practice in communities residing in migratory corridors and dispersal areas where there is maximum interaction between livestock, wildlife, and humans. However, this can be secure in communities where the government routinely issues buffer zone vaccinations as a strategy for safeguarding livestock and humans from zoonotic diseases from wildlife. In a similar vein, any wildlife tourist's activity involving the pursuit and distress of an untamed animal is detrimental.

For example, pursuing and cornering wild animals during trophy hunting causes several traumas to wildlife species. Similarly, wildlife viewing may also be distressing as some tour guides pursue what is termed tourists' preferred species for an extended time and distance for wildlife tourists to have a closer view of the animals. In many protected areas, this is a severe offense, but they do it because they are being paid extra money or for any other purpose. When the feral animals are agitated because of the reasons mentioned, their natural immunity also goes down, leading to disease acquisition, and ultimately the animal may perish.

Moreover, transmissions of disease-causing pathogens between wildlife and tourists depend on the type of activity the wildlife tourists indulge in; those activities involving intimate interactions with wildlife species are likely to maximize the change in transmission. For example, animal feeding in private protected areas, wildlife farms and ranches, and zoological gardens, where the wildlife attractions entertain animal feeding by permitting the wildlife tourists to feed the animal at a close distance where disease transmissions are likely to occur.

Also, the closer the genetic relationship to humans, the greater the likelihood of wildlife species exchanging disease-causing pathogens between humans and wildlife. Some wildlife species, such as primates, are closely related to humans, and therefore transmission of some diseases becomes easy. For example, the transmission of flu between humans, gorillas, and chimpanzees is easy because their genetic materials are close.

I can recall that in 2003, one of Tanzania's famous radiostations advertised that a baboon, from one of the Eastern African protected areas, was diagnosed with a sexually transmitted disease. Though it was not officially confirmed or criticized by the responsible conservation agency, in some cases, wildlife tourists are involved in unnecessarily close interactions with wildlife to the extent of acquiring or transmitting.

Imagine the situations where wildlife visitors attempt to bite or break a wild animal's carapace or offer water or a drink to a baboon. In the past, a tourist offered a beer to an African elephant to consume, but the story ended with the same tourist being killed after the elephant got drunk.

Encroachment of wildlife tourists and wildlife operators into wildlife habitats through wildlife activities through the construction of accommodation facilities may increase the interaction between wildlife and wildlife tourists, hence the flow of diseases. Wildlife tourism entails the construction of facilities within or near wildlife-protected areas.

In that case, some of the hotels are constructed within the protected landscapes, or wildlife migratory corridors, where the possibility of disease transmission is relatively high. Wildlife tourists interact with insects and baboons, among other things. In consumptive wildlife tourism, in particular spot fishing, hunting, or trophy hunting, wildlife tourists are allowed to handle and even consume the catch.

Capturing and returning wild animals to their natural habitat in sport tourism increases the chance of zoonotic disease transmission. Similarly, during hunting expeditions, consuming uncooked or partially cooked meat may result in disease transmission to wildlife tourists. Moreover, the captured wild animals, especially those intended to be used as pets, if untreated, are likely to transmit zoonotic diseases to the owner.

Changes in Wildlife Behavior

Wildlife tourism can affect the natural behavior of wildlife, such as aggressiveness, construction of wildlife infrastructure in corridors and migratory routes, stress on wildlife, e.g., many vehicles per animal, photographing, disease transmissions from tourists to wildlife, poaching of wildlife, mainly elephants and rhinoceros, construction of tarmac roads, e.g., Kruger National Park, and killing of visible and invisible biodiversity, e.g., driving off the road and over speeding.

Wildlife tourism may alter wildlife behavior when it is not properly controlled, and various wildlife tourism activities may inflict varied impacts on wildlife species. One of the common impacts of game observing or wildlife viewing is habituation when wildlife loses entirely a dread for humans or objects. The untamed animals' frequent innocuous encounters with humans or objects make them fearless of humans, causing them to lose their natural behaviors.

As a result, the animals spend most of their time closer to tourist-designated roads or accommodation facilities for food solicitation. In addition, wildlife tourists tend to feed or even share beverages with habituated wild animals. It feels enjoyable and engaging for wildlife travelers to feed wildlife in the natural environment, though it is dangerous to share food and beverages with wild animals. It makes untamed animals alter their natural behaviors by becoming wholly dependent on tourism for sustenance and water.

With time, the untamed animals lose the ability to seek sustenance by themselves and become petitioners. Feeding untamed animals artificially is a threat to their feeding behaviors and health, as it may lead to the utter loss of normal feeding behavior. The inability of wild animals to locate their natural sustenance animals may affect their health status by either making them overweight or emaciated.

In high tourism seasons, routine provision of food makes the animals overweight, while during the off-season, the low number of wildlife tourists nourishing the wild animals makes the animals slender. Unethical wildlife travelers and operators feed untamed animals while on vehicle safaris or in hunting cabins. During peak hunting seasons, tourist sites produce and accumulate enormous amounts of refuse.

The recyclable and unrecyclable materials generated necessitate special waste management and disposal approaches, as insufficient management and disposal have a negative impact on wildlife species. The biomass discarded by wildlife visitors has a significant impact on species' population densities and dynamics, reproductive success, behavioral adaptations and movements, and habitat utilization.

Moreover, the residues are poisonous to untamed animals. Spatial dispersion and the accumulation of sustenance normally determine species distribution and species richness in a certain area. Accumulated and indisposed food and non-food residues impact the population distribution of various species in the protected areas.

In some areas, an imbalanced population of certain species increases, making them conspicuous and vulnerable to predators and intruders. Unattended refuse also attracts animals, making them highly dependent on waste. As a result, the animal loses its instinct and becomes unable to forage naturally. Mismanaged recyclable waste attracts many animals to certain areas, thus accelerating human–wildlife conflicts and species aggregation.

However, animal activists unjustly accuse trophy hunting of causing immense waste production when other categories of tourism activities, such as mass tourism, can produce more recyclable and unrecyclable waste than trophy hunting. Furthermore, trophy hunting occurs seasonally, while mass tourism occurs throughout the year, where the number of mass tourists exceeds that of trophy hunters.

Due to the difference in the number of tourists and the periodicity of the activities, mass tourists and ecotourists produce more pollutants than trophy hunters. Unhealthy food distributed casually to wild animals, including tinned food, may have physiological impacts on wild animals by interfering with their reproductive systems. The failure of untamed animals to reproduce appropriately implicates the species' existence and viability.

Studies show that artificial food has several impacts on the reproductive behaviors of wild animals, such as male alligators having ovaries after a long exposure to agricultural farms feeding the fish and lions without conspicuous manes after a long time of feasting on the livestock. Viewing animals can drive them away, disrupt their grazing and reproductive sites, or acclimate them to the presence of humans.

To observe wildlife and take photos closely, wildlife travelers may drive away wild animals from reserves, increasing the risk of inbreeding them and further endangering the species. It disrupts family-specific relationships. Female animals' attendance at their offspring decreases when visitors are present. Females who remain with their children spend significantly less time nurturing and more time observing travelers.

There is also a risk that young people will not be recognized and will be more vulnerable to predatory attacks. Wildlife tourism, in particular trophy hunting, affects the appreciation and valuation of some species by causing them to be more famous and valued than others. Trophy hunters regard game species as the only valuable wildlife species in the world, which causes some protected areas or countries to be more visited than others.

Tanzania, for example, is the most famous destination for hunting buffaloes, leopards, and lions, and tour outfitters use these animals to promote their companies and hunting destinations to attract tourists. Also, wildlife tourism, directly or indirectly, affects the behaviors of game and non-game species, as it interferes with the grazing, spawning, and distribution patterns of the animals through gunfire, pursuit, and disturbances.

Trophy hunting intensifies human–wildlife conflicts because the targeted species tend to abandon the hunting grounds for safer places to prevent fatalities, disturbances, and injuries from recreational hunters. When fugitive or frequently pursued species migrate to areas closer to the human landscape, the incidence of human–wildlife conflicts may increase. Consequently, carnivores might attack domestic animals, while herbivores might plunder harvests and irritate humans.

Furthermore, the abrupt change in native habitat may impact the species' genetic diversity, reproduction, fecundity, and mortality rates due to their exposure to unfamiliar environments. For migratory birds, for example, fat deposition and storage are extremely important before distant migration. Trophy hunting significantly impacts fat deposition and storage in migrant birds because, during the trophy hunting season, birds become more agitated than in the off-season.

However, trophy hunting is not the only cause for animals to alter their behaviors. Changes in land use, climate change, exploitation, human-wildlife conflict, and tourism photography may all contribute to the change. Feeding animals by photographing visitors makes them highly dependent on artificial feed. Moreover, some animals abandon their grazing and reproductive areas to avoid noise and disturbances from travelers or development projects.

Worse, some animals may metamorphosize or alter their body structures and behaviors to avoid exploitation. For example, some African elephants go tuskless to avoid ivory poaching. Some change their activity to nocturnal or corpuscular to avoid poaching incidents at certain times of the day. Similarly, human-wildlife conflict alters the grazing behavior of some animals; for example, inveterate crop invaders become completely dependent on domesticated crops because they taste more palatable, nutritious, and digestible than vegetation and trees.

Population Dynamics of Wildlife

Wildlife tourism can have a variety of negative impacts on wildlife populations. Some of the impacts include an increase and a decrease in wildlife populations. Various wildlife tourism activities may have varying impacts on wildlife species. Some wildlife tourism activities, such as trophy hunting, affect the population structure by driving away individuals of certain game species because of uncontrolled murders.

In short, wildlife tourism may change the dynamics of wildlife populations through behavioral, psychological, habitat damage, anatomical changes, and human-wildlife conflicts. Wildlife tourism adversely affects the behavior of wildlife by interfering with their spawning, grazing, mating, and migration.

In the process of attaining exceptional features and behaviors from untamed animals, wildlife tourists may interfere with the reproduction, migration, and grazing patterns of wildlife, therefore affecting wildlife populations. Conducting destructive wildlife tourism activities during the breeding season and in the breeding or nesting areas affects wildlife breeding behaviors and activities.

Trophy hunting is one example of an activity that, when conducted during the reproductive season, may alter the breeding, grazing, and breeding behaviors of some wildlife species. As a result, there has been a reduction in the reproductive rates and success of some wildlife species.

Some individuals may forsake their reproductive sites and areas by migrating to other areas with minimal disturbances. Multiple studies corroborate that in East Africa, the reproduction rates of many carnivores have been extensively affected by wildlife tourism activities.

Environmental Degradation

In this book, environment means the natural habitat of wild flora and animals. The places where the natural animals and flora spend most of their time can be water, land, air, mountains, and trees, as well as other vital resources. The discussion on the environment can be very demanding, encompassing many pages of this chapter.

As long as the objective of this chapter is to emphasize the environmental consequences of wildlife tourism, the environmental discussion will only focus on a few components. Wildlife's environment is also referred to as a natural environment or natural habitat. Before we forget, sustainable wildlife tourism is beneficial to the natural environment, but wildlife tourism is not.

This is a very essential point to remember. That being said, wildlife tourism may cause infinite undesirable impacts on soil, water, air, and other constituents of the environment. The essential point to note here is that the extent of adverse impacts inflicted by wildlife tourism on protected areas may substantially differ between protected areas. That is, even in protected areas with a similar policy framework and under the same conservation agency, visitors may experience differing levels of adverse impacts from wildlife tourism.

Some protected areas may be more or less impacted than others. The important points to note from this perspective are that the many adverse impacts of wildlife tourism can be controlled only if all wildlife tourism operations taking place within the protected areas are properly planned and wildlife behaviors are strictly regulated and monitored according to the plan.

In summary, the adverse impacts of wildlife tourism are preventable and controllable. Because prevention is better than cure, all impacts must be avoided by implementing sustainability policies. Wildlife tourism has a destructive environmental impact that threatens the disappearance of keystone species and essential environmental services.

To survive, the livelihoods of global communities depend on environmental services; water, medicines, oxygen, and food are examples. In that context, protectiveness becomes preferable to reactiveness for conservationists and destination managers to prevent and reverse any environmental consequences of wildlife tourism.

Uncontrolled wildlife tourism is likely to cause land degradation, water contamination, deforestation, and air pollution. Wildlife travelers need local transit at their destination, particularly in developing countries where geographical scenes are exceptionally challenging and necessitate them to drive for hours to seek out animals with appropriate mementos.

Because most greenhouse gas emissions are caused by energy consumption, road transport emits the majority (74%). In this case, wildlife tourists emit greenhouse gases while hunting. Serrano-Bernardo et al. (2012) emphasized that hunters' mobility, mode of transport, the total number of kilometers traveled, days of journeys, and efficacy of travel equipment determine energy consumption and therefore the quantity of greenhouse gas emissions.

Additionally, wildlife visitors emit greenhouse gases through services and accommodations at the destination. The emissions are caused by the consumption of energy in the form of heat and electricity in the accommodation facilities.

Species Endangerment

Trophy hunting is the primary cause of both intentional and unintentional overhunting. Sometimes overhunting occurs when unethical trophy hunters join hands with unethical wildlife managers or hunting operating companies to kill more animals than specified in the hunting permit or hunting quota, while unintended overhunting results when the conservation agency issues more hunting quotas than the actual number of individuals and species in the hunting block. It happens due to a lack of sufficient information on species abundance and availability.

When the number of individuals killed exceeds the species' reproductive rates, overhunting becomes more disastrous. It usually drives a population of frequently hunted species to extinction. As a result, several species undergo extinction due to overhunting. Currently, some species are close to extinction due to the overhunting caused by trophy hunting.

In essence, overhunting has reduced the populations of African lions (*Panthera leo*) in Tanzania and Zimbabwe, as well as American cougars (*Felis concolor*) in America. It is important to note that resident hunting, wildlife capture, poaching, human–wildlife conflict, and live animal trade all result in overharvesting wildlife resources by removing animals from their natural environment.

If mismanaged, such harvesting results in an underpopulation of certain species and individuals. Poaching is the leading conservation challenge after human-wildlife conflicts in Africa. Ivory poaching, for example, significantly reduced the population of African elephants (*Loxodonta africana*) from 800,000 in the 1970s to 86,000 elephants in the 1990s. It equally reduced the population of black rhinoceros (*Diceros bicornis*) from 60,000 in the 1960s to 3000 in 2013.

Human-wildlife conflict has been reducing the wildlife population by causing needless wildlife deaths. A recent study estimated that the number of lions killed by humans because of human–carnivore conflicts were directly proportional to the number of livestock deaths in the Serengeti ecosystem.

Such killings are monotonous and attract urgent conservation attention because human–wildlife conflicts usually kill animals unselectively regardless of their ages, rarity, sexes, involvement in incidents, and conservation status. Furthermore, there are countless wildlife species and individuals confined within zoological gardens, which are off-natural ecosystems, as well as those involved in animal trafficking.

Impacts on the Natural Vegetation

Wildlife tourism may cause various adverse consequences for natural vegetation. Natural vegetation includes forests, woodlands, grasslands, and individual untamed plants. Natural vegetation is one of the most prominent assets for wildlife tourism. Wildlife visitors may have an impact on natural vegetation in a variety of ways, including overconsumption, stampeding, compaction, the introduction of foreign invasive species, biological piracy, and illicit off-taking, among others.

Degradation of natural vegetation may occur without the knowledge of wildlife tourists or intentionally, with or without the collaboration of wildlife operators and some unscrupulous conservationists. All in all, wildlife tourism's impacts on natural vegetation are imminent and avoidable. It is essential to cite some of the specific impacts of the ways wildlife tourism severely affects natural vegetation. Wildlife visitors have direct and indirect influences on the natural vegetation; one of them is deforestation.

It is the opposite of afforestation, but it simply means the illicit cutting or harvesting of natural vegetation, which can be forests, grasses, or long trees. In summary, deforestation is the illegal and legal exploitation of timber and non-timber forest products.

In that context, deforestation means the illegal harvesting of natural vegetation to augment the food and non-food demands of wildlife tourists. It implies wildlife tourism influences the illegal exploitation of natural vegetation, which is the primary attraction for wildlife tourists but also the main source of environmental services.

It is difficult to comprehend how wildlife visitors may be implicated in the illicit harvesting of natural vegetation. It may sound intriguing without any further explanation. I sometimes fail to comprehend if wildlife tourists know exactly the purchasing and procurement procedure for the construction materials for luxury tourism superstructures, including cruise ships, accommodation facilities, trains, automobiles, and airliners. If the answer to this concern is yes, then are they under local, national, regional, and international laws? If the answer is either I don't know or no, then wildlife tourists indirectly support the illicit exploitation of natural vegetation.

Wildlife travelers also consume natural foods, such as mushrooms and vegetables, harvested from protected areas with or without their knowledge. The purchase of mementos and carvings made from endangered wild plant species is linked to piracy among wildlife tourists. In other words, it is referred to as the illegal harvesting of non-timber natural vegetation.

Responsible wildlife travelers always search for correct information before purchasing anything at a destination to safeguard the environment, culture, and social aspects. As a result of irresponsibility, there is an immense loss of biological diversity due to uncontrollable wildlife tourist activities.

Furthermore, wildlife visitors are also significant sources of alien invasive species in natural ecosystems. Alien invasive species rank among the top three challenges to protected areas. Most of the invasive species are culturally, ecologically, and economically destructive. Ecologically, alien invasive species lack natural predators or adversaries, so they tend to have unlimited and robust growth to the extent of getting rid of many native species.

With time, the native species that are naturally alluring and socio-culturally essential to the communities may completely disappear. Wildlife visitors may introduce the alien invasive species with or without their knowledge, for example, by tossing exotic fruit seeds in protected areas. Strolling can spread germs through shoe soles.

The disappearance of native species of natural vegetation may also influence the abundance, diversity, and distribution of untamed animals. Similarly, the sociocultural identities of the neighboring communities may disappear as the native species that were used as food, medications, or construction materials have completely disappeared.

Moreover, the scenic features caused by the extinct vegetation may also disappear. The pertinent example in this matter may be Kitulo Plateau National Park in Tanzania, where blossoming natural vegetation is the primary attraction of the protected areas. The invasion of exotic species into these protected areas may be ecologically calamitous.

Source of Wildfires

Wildlife visitor activities are among the causes of conflagration in protected areas. Any conflagration out of control is termed a wildfire. In the dry season, protected areas have enormous depositions of desiccated, perished forests that are very susceptible to fire. That heightens the possibility of visitors starting wildfires accidentally or intentionally.

Wildlife visitors participate in a variety of activities in the protected area that cause wildfires. All these happen because wildlife visitors and tour operators do not sufficiently follow fire safety rules and protected areas' regulations that rigorously prohibit smoke of any description in the natural areas. When camping, use equipment that can easily cause wildfires, such as vehicles, generators, and gas cookers.

Camping is an adventure mostly enjoyed by some wildlife travelers. It entails the use of fire apparatus for cooking and illumination during the night. Improper control of the fire apparatus may cause fire to escape. Depending on the camping location, the escaped fire may be calamitous and hard to control due to the massive and prolonged deposition of the deadwood.

The conflagration is costly to control and wildlife, mostly affecting burrowing, elderly, ill, and juvenile animals. Additionally, leaving the fireplace unattended after camping and improperly disposing of cigarette stems after smoking. Failure to turn the fire off after camping is one of the most common causes of conflagration in protected areas. In some protected areas, it is legal to start a fire at the campsites, but some of the visitors regrettably neglect to turn it off themselves. Such a fire can have a detrimental impact on protected areas if it starts unnoticed.

It is the responsibility of tour operators and protected area administrators to remind visitors to properly respond to the fire. Moreover, the use of apparatus like vehicles that may malfunction and ignite fire is another cause of conflagration in the tourism area. Poorly maintained tourist vehicles may ignite fire in protected areas, causing conflagration.

Though these are rare cases, they could be one of the sources of confusion. Despite the sources, putting the conflagration off is an expensive, hazardous, and time-consuming process. Though it hardly affects the lives and property of wildlife travelers, it catastrophically affects protected resources. In that sense, collective efforts are essential from various institutions to rescue the scarcely protected resources.

Land Degradation

The land contains untamed animals, humans, and vegetation; it is home to many tourist attractions. It also absorbs the many disturbances caused by wildlife tourism operations. Wildlife visitors also use the land as a transportation and accommodation platform.

More importantly, the food and water consumed by wildlife tourism at the destination are sourced from the land. In brief, at least everyone now receives a clear picture of how wildlife tourism wholly depends on land resources and, unfortunately, how wildlife tourism operations may severely impact land from different aspects.

The wisdom of wildlife visitors, tour operators, and conservationists is absolutely important if we want to investigate the benefits of land forever. Contrary to that, wildlife tourism destinations are likely to confront recurrent soil erosion, compaction, mud sliding, and soil infertility. It is obvious that, like other forms of tourism, wildlife tourism also has five essential fundamental elements, including conveyance, accessibility, and accommodations.

It means that without these two main components, ecological tourism is insufficient. Wildlife tourism, on the other hand, entails the development of accommodation facilities and transport infrastructure in wildlife-protected areas. The problem is not the construction of these essential facilities and infrastructures, but rather the materials used, the technology and procedures involved, and the locations where they are built.

The lecture on general management planning demonstrates that various areas within the protected areas may have differing levels of resilience against the same pressure of development intervention. That is why we critically need management zones during planning and implementation. That being said, the negligent construction of tourism mega-infrastructures in protected areas may threaten ecologically sensitive areas of land.

Wildlife tourism typically consumes virgin areas of land where there is a lack of information on resilience and sensitivity. Moreover, the tourist transport facilities, including vehicles and flights, hardly correspond to the actual bearing capacity of the protected areas. As in many cases, the number of tourist vehicles traversing the protected areas increases as drivers pursue the magnificent species in the protected areas without considering the bearing capacity of the automobiles and tourists.

Moreover, wildlife tourism attracts the designation and construction of tourist routes that involve the eradication of natural vegetation and the thorough excavation of soil. These lead to the destruction of the microhabitat of terrestrial and subterranean organisms. Heightened casual construction of tourism facilities and infrastructure causes soil erosion and land and earth sliding in ecologically sensitive areas. As a result of the construction pressure on the land, the exertion of severe pressure on the ecosystem and local resources, and the scenic features of the protected areas.

For example, there has been high pressure along the coastlines as everyone attempts to construct enticing tourism facilities on the shore. According to the UN Census, 60% of people worldwide live along the coast. The direct impact on natural resources in the provision of tourist facilities can be caused by the use of land for accommodation and other infrastructure provisions and the use of building materials.

Water Pollution

Water-based wildlife tourism activities generate enormous repercussions on water, which include aquatic habitats and the supply of drinking water for animals and people. The lack of adequate sanitation facilities and proper waste management systems in some protected areas results in casual waste disposal, as some wastes are disposed of in wetlands, rivers, lakes, and oceans, contaminating wildlife, livestock, and people who usually fetch drinking water from them.

The discharge of dangerous inorganic chemicals into aquatic environments impacts animals, people, plants, cattle, and water-based tourist operations. Wildlife tourism activities can discharge sewage into the aquatic environment in a variety of ways, both directly and indirectly. Some of the most common methods involve lodging and leisure facilities.

The construction of hotels, leisure facilities, and other amenities for wildlife visitors typically leads to increased sewage contamination. Wastewater is impacted by natural vegetation and animals. Certain wildlife tourism facilities are created in the lake or ocean when relevant knowledge of the specific location's sensitivity and resilience is lacking. After construction, the long-term harmful impact of building materials and technology on the aquatic system is frequently overlooked.

Furthermore, some of the lodging establishments are erected without undertaking environmental impact studies or establishing mitigation strategies to lessen certain construction consequences. Surprisingly, everyone realizes that sewage pollution in aquatic environments harms breeding grounds for fish and other aquatic species, like turtles and crocodiles.

The destruction or interference with wildlife breeding habitats may have enormous, catastrophic repercussions on the number and variety of water species. Additionally, the development of demand for seafood and fish by wildlife visitors may boost aquaculture production in the natural aquatic system to fulfill the food requirements of wildlife tourists.

Furthermore, intense artificial agriculture and sewage runoff from tourist facilities increase algae development in the rivers, seas, and lakes that cover the filter-feeding corals, hampering their capacity to live in their natural ecology. The overabundance of nutrients in bodies of water leads to eutrophication. Several incidents highlight the consequences of excessive nutrients in water bodies. For example, large flamingo fatalities in Lake Manyara and Bogoria, Kenya, occurred around 2006.

Moreover, eutrophication may harm the health of people and animals, rendering the bodies of water worthless for any water-based tourist activity. In similar situations, the release of toxic chemicals such as chlorine into wetlands has a severe impact on biodiversity, hampering any water-based wildlife tourism activities. Moreover, wildlife visitors barely comprehend and recognize how much water is in the meals they eat, the toilets they flush, the drinks they drink, and so on.

It is vital to highlight that whatever quantity of water they use is still a component of the local, national, and regional hydrological and ecological systems. Being aware of their impact may make them environmentally responsible.

In some nations, animal attractions are in agriculturally neglected regions where water and nutrients are scarce. In other countries, there is a fierce rivalry over water resources between residents and wildlife visitors, as well as between wildlife and cattle. Despite the great water shortage, wildlife tourism businesses often overuse water resources for hotels, swimming pools, and golf courses, as well as personal water usage by guests.

This may lead to water shortages, the deterioration of water sources, and a greater amount of wastewater. Water shortages are disastrous in many drier places, such as coastal regions. This could lead to water problems due to the hot environment and the propensity of visitors to drink up to 440 liters a day, which is 23 more than the local population.

Realistically, such large water consumption and disparity may appear unjust and unpleasant, as in the East African area, a resident or local individual may take roughly an average of 19 l of water in a day. It is just a water dispute. What is the purpose of being cheerful while your neighbor's family next door is dying of thirst? What makes you joyful when your neighbor's cattle are dying of thirst?

While wildlife tourists spend the same water that might conserve natural resources, their neighbors, families, and cattle are swimming, and they use it for what you call recreation. It is one of the biggest failures and worst repercussions of wildlife tourism, as it entirely fails to balance the demands of wildlife visitors, people, animals, and livestock.

Water is a very vital resource. I recall reading a WHO study in 2010 stating that water kills more people than wars, and there is also a possibility that the next global war may be sparked by a lack of water resources. Water is extremely vital for wildlife; it is one of the predictors of human–animal conflicts.

I still remember the episodes in 2017 when I was completing my research in a village named Balili near Serengeti National Park when an African elephant had to damage the water tap to get water during the dry season.

Socio-Economic Impacts

Poor Living Conditions

Foreign investors own the majority of ecotourism ventures, such as Tarangire Sopa Lodge and Lake Manyara Tree Lodge, among others. The majority of earnings are diverted into the wallets of investors instead of investing in the local economy or environmental preservation. Because of poor pay, a small number of local people who work in the economy are unable to live in tourist zones. The local people's lack of economic prospects also prevents them from destroying the environment as a means of subsistence.

Mismanagement of Natural Resources

Governments frequently lack the commitment or competence to manage ecotourism areas properly. The rules for environmental protection may be loosely defined, expensive to establish, difficult to enforce, and unclear in efficacy. Government regulatory agencies, as political authorities, are tasked with making decisions that allocate funds to politically desirable but ecologically counterproductive initiatives.

Commodification of Culture

Wildlife tourism may transform the culture into a commodity. The monetization of plants may take away social value and lead to overproduction inside protected areas as local people, their images, and their culture are converted into commodities for commerce. The formation of protected zones may create unequal survival conditions and deprive the people of their customary usage of land and natural resources for their socio-economic development. Wildlife tourism forces locals to compete for cultural expression while being watched by visitors.

Displacement of Local People

Wildlife tourism may displace indigenous people from their natural locations to either expand the protected areas or create infrastructure or a superstructure for wildlife visitors. The Masai narrative is one of the most frequently referenced in East Africa.

About 70% of protected areas in East Africa are on the Masai ethnic group's ancestral grounds. Protected areas took advantage of their ignorance about the issue and stole significant portions of grazing land, putting them in danger.

Socio-Economic Life

Formation, upgrading and resizing of protected areas make many locals people lose their homes while receiving little or no compensations. Wildlife tourism misplaces Indigenous people in marginal locations with severe weather, poor soils, a lack of water, and highly afflicted areas. Such agriculturally neglected regions rarely sustain any agricultural activity, leaving local inhabitants even more destitute.

Despite their loss of property, wildlife tourism's employment prospects overlook the local community and favor other individuals, notably foreigners and those with higher education. Furthermore, the investors in this region are not locals and have not contributed earnings back to the local economy.

In rare situations, protected zones might be formed without the consultation of local people. Wildlife tourism, as a multinational business, barely empowers local people; rather, it exploits them. It needs to be community-oriented if it is meaningful and sustainable. The Maasai tribe has been forced to leave their land to make room for tourists and trophy hunting.

Local Scarcity of Resources

Tourism may impose significant demand on local resources like energy, food, and other basic commodities that may already be in low supply. Because of the seasonal aspect of the business, many places have 10 times more population in the peak season than in the low season.

To deal with the high demand for local resources, there will be an increase in the production of local resources by adopting inorganic agriculture and sophisticated modern technology to supply the expanding need for local resources. This frequently leads to environmental degradation.

The negative environmental repercussions of tourism are enormous. They include the loss of local natural resources, pollution, and waste concerns. Tourism typically puts strain on natural resources via overconsumption, sometimes in locations where resources are already limited.

Tourism puts a huge burden on local land usage, which may lead to soil erosion, increased pollution, natural habitat loss, and additional strain on endangered species. These consequences may eventually damage the natural resources on which tourism itself relies.

According to surveys, vacationers may drink approximately six times more on holiday than at home. However, some of these consumptions may wind up straight in the bins of hotel kitchens and restaurants, which are responsible for 12% of worldwide food waste. Such meals would have assisted the inhabitants of the area.

Moreover, the energy consumption of wildlife visitors on vacation is substantially greater than at home, including consumption for heated pools, local transit, or air conditioning, producing energy crises at certain sites. In developing countries, electricity and clean energy are always lacking.

Part III
Sustainable Tourism Practices

Chapter 5
Sustainable Wildlife Tourism

Abstract This chapter presents and illustrates the concept of sustainability in wildlife tourism. Sustainable wildlife tourism is a concept embracing ethical executions of wildlife tourist operations from attraction to global levels. It considers economic development, community well-being, and environmental conservation in all wildlife tourist activities. It is not an obligation of the governments solely but of all wildlife tourism stakeholders, including tourism businesses, as well as local communities, and wildlife tourists must collaborate, practicing and promoting responsible behaviors. With sustainable wildlife tourism, it is the responsibility of every stakeholder to enforce legislation and support national, regional, and global wildlife conservation-orientated development. However, the wildlife tourist destination should integrate sustainability principles into policy and operations to support the socioeconomic, cultural, and environmental responsibility behaviors among wildlife tourism stakeholders. In wildlife tourism, sustainability increases the intelligent use of wildlife resources while considering the requirements of current and future generations. Sustainable wildlife tourism reflects the suggestions from the Brundtland Report, which stipulated the main principles revolving around societal equity, environmental conservation, and equitable relationships between generations. The sustainability principles regulate the implementation of ethical tourism operations to reduce species endangerment and ecological disturbance, enhance community development, respect indigenous culture, and promote community involvement in wildlife tourism at various levels. Adoption of the social sustainability concept through cultural preservation, community participation, and fair cost-benefit sharing schemes could improve the involvement of the local community in not only wildlife conservation but also wildlife tourism. Sufficient and timely involvement of local communities in decision-making fosters a sense of inclusion and ownership and protects and supports cultural traditions and knowledge against tourism exploitation. Responsible tourism might equally help local economies by developing local economies while supporting local suppliers, promoting small enterprises, and preventing economic leakage.

Keywords Sustainability · Environmental impact · Social equity · Economic resilience · Community-based tourism · Ethical wildlife management

A. A. Mamboleo, *Sustainable Wildlife Tourism*,
https://doi.org/10.1007/978-3-031-97795-4_5

Sustainability of Wildlife Tourism

Sustainability is the idea adopted from sustainable development in the Brundtland report, Our Shared Destiny, presented by the World Commission on Environment and Development (WCED). When I was young, I was intrigued by the pronunciation of Rio de Janeiro, one of the most renowned Brazilian cities.

I came up with the name every time I listened to the national radio, Radio Tanzania Dar es Salaam, now Tanzania Broadcasting Corporation. At the end of the 1980s and early 1990s, world leaders convened in Brazil to discuss strategies for revising the terrible environmental destruction and excessive poverty highlighted by the report.

Among numerous other things, the paper proposed the notion of sustainable development and how it might be collaboratively and simply realized. According to the research, sustainable development implies the capacity to use natural resources without compromising the ability of future generations to access them.

Between the 1970s and 1980s, the coinage and introduction of sustainable development evolved after unending environmental degradation, characterized by bloody poaching and deforestation. Similarly, extreme poverty worried some activists to the point of doubting the urgent measures to reduce the existing gap between the poor and the affluent.

After an investigation into the causes of environmental degradation, the WECD aimed to link social fairness, economic development, and environmental challenges, which led to the present policy solutions that combine all three domains.

The Brundtland report anticipated that if developed countries maintain the same economic development pathway of resource overexploitation and developing countries ignore alternative pathways to the development of overdependence on natural resources, a massive climate and resource crisis may occur in the world.

As the ghosts of climate change and crises still bite our planet, the global ecosystems are overstressed, and poverty is rising. As one of the recommendations, the Brundtland report proposes that any development intervention should address social, economic, and ecological issues, and these came to be the key paths and pillars of sustainable development.

In brief, the population also grew increasingly conscious that there was an element of uncertainty and danger regarding the influence of a variety of human activities on global surroundings. To mitigate these difficulties, fundamental adjustments were necessary in our way of life. This awareness sparked the notion of sustainable development.

The research evidences several concerns to be addressed and remedied promptly, including intergenerational equality, intragenerational equity, public engagement, and environmental protection. Sustainable development, from the perspective of intergenerational fairness, should equally reflect the demands of present and future generations.

Studies show that sustainable development should consider both current and future needs. What is provided for later generations should also be available for the present generation. It is our ethical duty and responsibility to pass on the resources to future generations exactly as we inherited them from previous generations.

Moreover, the research highlights that the present generation should carefully and intelligently employ the resources in a manner that does not contradict the quality and quantity of the same resources for the upcoming generations. Because if past generations overused or deteriorated the already-inherited resources, the current generation would not be able to access and enjoy them now.

Similarly, intra-generational equity may remove inequities in accessing resources across residents and countries within the generation. Inequalities produce unequal access and resources for socio-economic growth and environmental conservation among individuals locally, regionally, and internationally. As a result, the divide between the affluent and the poor is expanding.

The Brundtland report emphasizes social justice, social empowerment, and poverty reduction to save vulnerable populations. Special strategies for poverty alleviation, social empowerment, and justice are needed to close the gap between the poor and affluent, not only benefiting the powerful or rich individuals or countries.

Several studies affirm that individuals living inside or adjacent to unique resources are the ones living in severe poverty and are uneducated. Social and human rights are often neglected by powerful and affluent people who readily exploit resources.

Additionally, public involvement is a vital component in sustainable development, meaning that every individual plays a big role and that every community has to make its own choices instead of having its decisions made by people outside the locality. Participation in decision-making promotes a feeling of owning and putting into effect their own choices.

Communities have to decide on their development destiny since they know what they want and have adequate traditional knowledge to address most of their development issues and problems. Due to socio-economic transition, glocalization, and globalization, many communities are aware enough to cope with their development concerns. It is time for governments to transition from centralization to decentralization to create opportunities for communities to decide their destiny.

Moreover, the Brundtland report highlights environmental protection as a basic part of any sustainable development. Since socio-economic growth entails the usage and destruction of environmental resources, environmental conservation should be a fundamental component of economic development. In essence, economic progress without environmental protection is no longer acceptable.

In a nutshell, sustainable development is best suited to accept the sustainability of any socio-economic project as a guiding principle towards cleaner and more responsible investment. Taking into account all the components indicated in the Brundtland report may lead to well-articulated socio-economic initiatives that are not just cleaner and greener but also responsible for communities.

Clarifications and recommendations from the Brundtland report led many disciplines and development partners to realize that sustainable development is a three-dimensional entity, requiring the incorporation of three pillars or principles, including environmental, community, and economic considerations. These have since become the key ideas or pillars of sustainable development.

Using this knowledge, several disciplines began operating sustainably, even prefixing their term "sustainable" to demonstrate an actual embrace of sustainable development ideas. It paved the way for terms such as sustainable agriculture, sustainable mining, sustainable forest protection, and sustainable fisheries, among others. Similarly, the tourism industry has embraced what is currently termed sustainable tourism.

Sustainable tourism was adopted to replace the detrimental repercussions of mass tourism, which will be examined more in the following chapter. Since tourism is so widespread, the sustainability concept extends to its typologies, including sustainable ecotourism and sustainable wildlife tourism, among others.

However, the purpose of this book is to place a greater emphasis on sustainable wildlife tourism than other types of sustainable tourism. Sustainable tourism appears to be one of the keys to minimizing the detrimental repercussions of wildlife tourism. Integrating wildlife tourism with the ideology of sustainable development is incredibly significant, as it may allow communities to engage and profit proportionally from wildlife tourism.

Sustainable Wildlife Tourism

According to the World Tourism Organization, sustainable tourism is defined as tourism that meets the needs of current visitors and host regions while also protecting and enhancing opportunities for the future. Sustainable tourism, in its purest definition, is an industry that strives to have a modest effect on the environment and local culture while also helping to produce money, employment, and the protection of local ecosystems.

It is responsible tourism that is environmentally, economically, and culturally attentive. In such circumstances, sustainable tourism operations have little influence on the ecology and culture of the host community. It takes into account the location's environmental and cultural sensitivity.

The act of exploiting natural resources for wildlife tourism sensibly without sacrificing the capacity of the present and future generations to use them is known as sustainable wildlife tourism, and it involves adopting the principles of sustainable tourism. In other words, it is a subset of sustainable tourism that follows all the principles of sustainable development.

The definition is so broad and increases the friendly and responsible utilization of wildlife resources in a thoughtful manner that benefits both the present and future generations without hurting the wildlife or damaging the local culture while increasing profits. Therefore, sustainable wildlife tourism takes on the rights and abilities of the present generations to reap the advantages of wildlife tourism.

Once wildlife tourism empowers the current generation and responsibly shares all the benefits, it creates awareness among tourism stakeholders about the importance of conserving wildlife resources, which are the heart of wildlife tourism, but it also creates a conducive atmosphere for wildlife tourism development.

This book emphasizes the importance of regulating wildlife tourism activities, not only as a way of limiting their detrimental environmental, social, and economic repercussions but also as a way of promoting their good sides. The integration of wildlife tourism operations with the pillars of sustainable development is one of the techniques for boosting the lifetime and productivity of wildlife tourism.

It is fair to say that the pillars of sustainable tourism development are global policies for tourism sustainability, which is why every destination, attraction, region, or country is required to align their tourist operations with them in the long term. In the 1970s and 1980s, horrible environmental degradation and economic and social catastrophe events happened internationally, and mass tourism was one of those perpetrators.

A large number of people visited the same tourist attraction at the same time. Such visitations led to regional and temporal polarization of tourist activities, which translated into tremendous repercussions for wildlife habitats and animal habits. Mass tourism scarcely addresses the carrying capacity and sensitivity of the delicate biological locations in both land and marine protected zones.

The Brundtland study was an eye-opener to many specialists, including tourism professionals, as it led to the adoption of alternative forms of tourism to safeguard vulnerable ecosystems and species. Alternative tourism demands that a few visitors visit the destination at the same time. It examines the ecological, social, and economic carrying capacity of the tourist attractions, which are vital in protecting the destination from poor resilience.

It resulted in the development of many tourist ideas and typologies, including ecotourism, nature-based tourism, medical tourism, geotourism, adventure tourism, and cultural tourism, among others. Massification of tourism is the notion of tourism that entertains a large number of travelers to visit a specific tourist attraction or location at one time. It is sometimes termed conventional tourism.

According to literature, mass tourism was first conducted in England before expanding to the rest of Europe in the nineteenth century. Mass tourism is the symbol and starting point of travel democratization. The emergence of mass tourism has been primarily caused by an increase in free time available for leisure activities, disposable money, and technical improvements.

Mass tourism occurs as a form of vacation, and it still takes place in certain locations. Some of the drawbacks of mass tourism include air pollution from tailpipes, hotels, and industry emissions, as well as physical damage to tourist attractions caused by trampling.

Also, mass tourism causes soil compaction due to the concentration of visitors in the same spot, as well as erosion in small areas due to the construction of tourist superstructures and infrastructure in natural regions. Mass tourism entails the development of large-scale tourist hotels without any concern for the ecological integrity of sensitive regions or undertaking environmental impact assessments to fulfill an inflated demand from mass visitors.

Moreover, soil degradation due to the construction of tourist facilities and vehicle movements, damage to vegetation due to trampling, water contamination via sewage or fuel spills, and debris from pleasure boats are all examples. Similarly, water contamination caused by hotel and restaurant dumping of human waste, casual littering in natural areas, and disturbance of animal or wildlife habitats are all examples of mass tourism.

Mass tourism supports an increase in social crimes such as drug misuse, prostitution, wildlife crimes, and excessive cultural commercialization. The consequences are especially obvious for young tourist places with inadequate resilience to mass visitation. More tragically, some individuals may even forsake other producing sectors, including agriculture and fishing, for mass tourist activities.

Massification to Alternative Tourism

Alternative tourism is the act of visiting a location with a relatively small number of visitors visiting a destination at one time and giving greater touch with the locals with respect and sensitivity towards the environment. It also enables pleasant contact between the inhabitants and the visitors. It has numerous forms, such as adventure tourism, ecotourism, cultural tourism, dark tourism, agritourism, medical tourism, sustainable tourism, and nature tourism.

The principal method to deal with the negative repercussions of mass tourism was to develop new tourist products, notably alternative forms of tourism. Environmentalists recommend educating visitors to be more conscious and respectful of the environment and its values. Tourists are growing less and less interested in group travel and want a tailored product with a high degree of novelty.

The problem is that many poor nations are unwilling to have ecologically responsible attractions and trip packages because tourism is a key source of income. The unfavorable environmental implications of mass tourist activities have led to a concentration on alternate kinds of tourism. It refers to an alternative to mass tourism.

Nonetheless, mass tourism attractions continue to exist and draw thousands of visitors each year. Some writers suggest that alternative tourism does not have to be an alternative to mass tourism but rather a conclusion or a path to progress toward. Alternative kinds of tourism are one of the new types of tourism that are emerging nowadays.

Mass tourism harms the environment, and businesses are developing alternative forms that are more environmentally responsible and sustainable. To reduce negative repercussions, the new forms control the number of tourists. Alternative tourism aims to reduce tourism's environmental costs while maximizing benefits.

Alternative tourism has several key features, including a limited number of visitors visiting a location at a specific time, small-scale lodging, gradual and regulated growth, local decision-making, and fuel-efficient transit. Alternative tourism is particular interest tourism (culture, event, adventure, etc.) or responsible tourism (natural tourism such as tourism).

Such tourist destinations emphasize increased engagement and understanding between hosts and guests, as well as between visitors and nature. Thus, alternative tourism may be considered synonymous with the notion of sustainable tourism. To distinguish these many sorts of tourism might be challenging.

In such a scenario, mass tourism is less sustainable than alternative tourism due to environmental costs. Alternative tourism obeys the ideals of sustainability by engaging environmentally conscious wildlife visitors. The tourism industry has undergone a significant transformation, particularly in terms of tourism consumption and production.

In truth, mass tourism is stagnant, even dropping, while alternative types of tourism meet the rising desire for responsible wildlife visitors and a fast increase. In mass tourism, the buying power is in the hands of producers, whereas in alternative tourism, the purchasing power is in the hands of customers who are also ecologically aware.

Alternative types of tourism are viewed as a replacement for mass tourism or the cure for the social, economic, and environmental repercussions imposed by mass tourism. These types of tourism are frequently guided by sustainability ideals. In most situations, travelers and tour operators are aware of the effects that may be produced by visiting sensitive sites.

That makes everyone accountable for protecting the environment and supporting the livelihood of local people. On the other hand, the Brundtland report improved visitors' knowledge about their ecological footprints, ultimately changing them into environmentally conscious tourists.

Sustainable wildlife tourism is a kind of alternative tourism that utilizes the principle of sustainability. It combines the three pillars of sustainable development recommended by the WECD. Wildlife tour operators, tour guides, or protected areas barely adopt all the pillars freely without any advice or order from governments.

Sustainable wildlife tourism is typically used to guide the country towards sustainable development. Each country has to turn global sustainability policies into national laws and regulations. More importantly, wildlife operators and protected areas may further integrate the ideas into their laws and regulations.

Many nations have already implemented, entirely or partly, sustainability principles to govern wildlife tourism activities. However, weak and poor law enforcement and education for tour operators and passengers are still concerns. Many wildlife tour operators still neglect the principles, and what they worry about is maximizing the profits of their enterprises while ignoring environmental protection and assisting local communities development.

Compliance with the principles of sustainability is crucial, as it helps in servicing the unique and endangered wildlife species that are on the verge of vanishing since wildlife species are not only significant for wildlife tourism; they also have many things to provide internationally. Currently, there are no other options than to protect animal species.

Wildlife resource utilization in wildlife tourism may vary depending on the protected regions, national legislation, and type of resource. Some protected areas may allow direct or indirect exploitation of animal resources; others may limit the use of wildlife to specific periods or throughout the year. Moreover, certain governments or protected areas may restrict tourist activity.

Some protected regions may designate separate tourist zones for resource utilization. Based on the factors described above, wildlife tourism may be consumptive or non-consumptive. Consumptive wildlife utilization refers to a sort of wildlife utilization that involves the taking off of wildlife resources. This type of wildlife utilization includes tourist hunting, resident hunting, and live animal trading.

Sustainability Principles of Wildlife Tourism

Sustainable wildlife tourism's main concepts, or pillars, are technically derived from sustainable tourism. It is the same idea that the World Tourism Organisation uses to define sustainable tourism. Generally, wildlife tourism still adopts some ideas from mainstream tourism. To move it to sustainable wildlife tourism, a lot of regulatory, legal, and infrastructural adjustments have to take place at viable sites. Wildlife visitors and attraction owners' perceptions of its appropriateness and relevance are some of the factors that require urgent adjustments.

Sustainable wildlife tourism is considered to be a quality type of tourism that entertains a limited but highly paid audience. It could have an impact on destination tourism and economic policy. As a result, many places may be reluctant to embrace and implement sustainable tourism concepts. That is to say, sustainable wildlife tourism may be financially and technically demanding, resulting in a partial adoption of a few principles of sustainable wildlife tourism.

Sustainable wildlife tourism, in contrast to traditional wildlife tourism, takes into account social, economic, and ecological sustainability at the local, regional, and international levels. The evolution of conventional wildlife tourism into sustainable wildlife tourism is a process that must, in the end, meet most of the social, ecological, and economic standards as stated by the World Tourism Organisation.

Environmental Sustainability

In the preceding chapters, I now and then stressed the need for wildlife tourism to be managed in such a way that it protects the integrity and elements of wildlife's existing natural habitat. The regions that have adequate knowledge of their fragility, constituents, and ability to resist the demand from wildlife tourism are absent. Wildlife tourism should sustain and safeguard all ecological processes, natural resources, and biodiversity.

Wildlife tourism operations should respect the natural processes that include river sources, breeding or nesting areas, migratory corridors, and migration activities. Wildlife activities that contravene natural processes become damaging and irresponsible.

Water-based wildlife activities at fish spawning sites interfere with the fish's spawning activities. The construction of tourist amenities and infrastructure in the migratory corridors, nesting regions, and dispersion zones harms natural processes.

Mfuwe Lodge in Zambia depicts the circumstances under which the lodge was built in a region where African elephants frequently went to eat wild mangoes. When elephants visited the following days to consume their natural mangoes, the trees were not there.

After the extended quest for mangoes, the elephants ended up at the reception counter, but the lodger was in the lodge. An unending number of similar cases exist, particularly in African settings, where the installation of tourism amenities among the natural landscapes is a privilege.

Unfortunately, such projects proceed without addressing biological processes, completing an environmental impact assessment to evaluate any possible repercussions, or creating suitable mitigation measures. Such reckless endeavors may barely upgrade wildlife tourist locations, wildlife tourists, or wildlife tour operators to sustainable kinds of wildlife tourism.

Ecological or environmental sustainability requires sustainable wildlife tourists, wildlife tour operators, and destinations to respect wildlife habitat, species, and the natural environment before the construction of any tourism facility, whether hotels, restaurants, or man-made attractions.

The welfare of wildlife comes first. However, each section has specific responsibilities in terms of ecological processes and biodiversity conservation. Wildlife tourism attractions or protected areas are accountable for proactive planning and installing adequate policy and regulatory frameworks for the smooth execution of sustainable wildlife activities. Governments need to ensure that laws and regulations are not only aligned with local, regional, and national tourist strategies but also with the overarching goals of sustainable tourism development.

The laws for wildlife operators and visitors to follow while at the wildlife sites are vital. The governments should establish code practices for wildlife tourism at all levels, including destination or protected areas, individual visitors, and wildlife tour operators. The recommendations for effect assessment, impact auditing, mitigation measures, and monitoring should be made at all levels for any place to move to sustainable wildlife tourism.

In Chap. 3, biodiversity protection is mentioned as one of the benefits of wildlife tourism. It indicates that protecting biodiversity alone does not make a wildlife tourist destination or tourism industry a viable entity, but rather the achievement of all criteria.

In other words, even traditional wildlife tourism occasionally helps to preserve biodiversity. Sustainable wildlife destinations and wildlife tour operators should conduct baseline environmental impact assessments before, during, and after the construction of wildlife tourism facilities.

Instead of waiting for the worst symptoms and signals, responsible organizations should always be proactive by understanding the possible repercussions of development initiatives before their actual manifestations. Moreover, self-monitoring and evaluation of wildlife tourist facilities is the next important characteristic of a sustainable entity, as they do not always wait for the relevant authority to audit their operations but collaborate sufficiently with other stakeholders to ensure effective monitoring, auditing, reporting, and mitigation of any impacts. In the case of wildlife tour operators, their conformance to professional and commercial ethics is vital.

The operators ought to be environmentally responsible and encourage responsible practices among wildlife visitors. As entities that spend most of their time and work closely with wildlife tourists, ethical wildlife tour operators should develop environmental awareness education for wildlife tourists. Responsible tour operators should prohibit any irresponsible activity, such as casual littering, feeding wild animals, poaching, off-road driving, chasing the animals, wildlife trafficking, and illness transfer.

Moreover, wildlife tour operators must dissuade wildlife visitors from buying food and souvenirs derived from endangered animals, including elephants, crocodiles, black rhinoceroses, and green turtles. Sustainable wildlife tourism demands that wildlife tourism stakeholders, directly and indirectly, perform research on the environmental or ecological implications of wildlife tourism on the environment.

The study is a technique for discovering the real consequences of wildlife facilities and tourism on ecological processes and biodiversity. That is to say, wildlife stakeholders may directly undertake research or indirectly give resources to research institutes to perform studies on environmental challenges. Furthermore, studies enable wildlife travelers to look for ethical wildlife operators and tourism sites.

In the tourist industry, several international and local wildlife tour operator firms employ greenwashing tactics by stating that they are ecologically friendly while engaging in various wildlife and social crimes. In such circumstances, ethical wildlife tourists perform simple research to check the validity of possible wildlife tour providers. Responsible visitors are also interested in discovering whether their payments go to conservation or communities.

Additionally, some wildlife visitors undertake easy research before purchasing any food or souvenirs to know the origin and components of the items. It is a means for ethical wildlife visitors to prevent local or international illicit wildlife commerce and assist wildlife conservation projects at local and international scales. Wildlife visitors may also question the recipes and sources of food they eat and shoes they purchase if they are neither from endangered species of animals nor acquired illegally.

Also, wildlife visitors may study the building sites and materials used for projects as a means of preserving endangered species and ecological processes. Tourist hotels situated near water sources, migratory routes, and breeding habitats may deter responsible travelers from utilizing such amenities. Furthermore, building materials from endangered species, exploiting water resources, and using poor waste management systems may dissuade responsible travelers from choosing particular tourist facilities.

However, the direct and indirect engagement of wildlife tour operators may motivate responsible visitors to pick a specific wildlife tour operator over another. In short, responsible wildlife tour operators, wildlife tourism attractions, and wildlife tourists all have to become ecologically responsible for preserving biological processes and biodiversity in protected and unprotected areas. The evolution of conventional wildlife tourism to sustainable wildlife tourism depends on each portion adhering to environmental or ecological sustainability methods.

Social Sustainability

Social sustainability is likewise viewed as a social pillar of sustainable tourist development or social equality. Quantifying social fairness, however, is difficult because it is typically assessed by employment rates, social infrastructure, empowerment, housing, food, and educational services, among other things. Sustainable wildlife tourism takes social responsibility more seriously than traditional wildlife tourism.

The degree to which sustainable wildlife tourism participates in social duties is practically greater than conventional wildlife tourism. Social responsibility encompasses various facets of communities. Sustainable wildlife tourism respects communities by returning a specific amount of wildlife benefit to them. It strengthens local populations by involving them directly and indirectly in wildlife tourism. Purchasing locally created items and utilizing local services by wildlife visitors make local people economically independent, which is also a significant component for increasing formal and informal job prospects.

As stakeholders impacted by wildlife tourism, their participation in making decisions on all topics touching their lives is crucial. The communities should engage in policy formulation, the selection of wildlife investors, the selection and ideas of advantages they desire from wildlife tourism, and the requirement of employing traditional knowledge for wildlife tourism.

Sustainable wildlife tourism conserves and respects the constructed and existing cultural heritage and traditional values of the communities since it is these values and cultural heritages that ensure the survival of wildlife resources, which are the major assets for wildlife tourism. Therefore, wildlife visitors and operators should respect their efforts to protect such cultural heritage.

Also, sustainable wildlife tourism adds to intercultural understanding and tolerance. Wildlife tour operators, location managers, and guides should support cross-cultural understanding and tolerance between locals and wildlife tourists instead of fueling misconceptions between hosts and guests. Wildlife visitors must raise awareness to properly understand the importance of respecting local culture and traditional knowledge. Wildlife visitors should respect local culture and customs while visiting wildlife regions enhanced by cultural influences.

For example, in several protected areas, some spots have been designated as worship sites for adjacent populations that are also regarded as tourist attractions within the protected area. Adhering to all local norms related to worshipping locations during the visit is crucial to supporting and respecting local culture, but it may also be a positive example of cultural tolerance for wildlife visitors.

Similarly, local groups view specific waterfalls, mountains, and rivers as hallowed locations. In the previous pages, the cultural importance of waterfalls and mountains was discussed. Wildlife visitors may visit such locations for various wildlife tourism experiences, but it is equally crucial that they clearly grasp and genuinely respect the underlying cultural values of the locals.

Sustainable wildlife tourism prohibits the over- commercialization of culture since it makes specific cultures unpleasant after long-term exposure to wildlife visitors. In the previous chapter, I attempted to show how wildlife tourism may over-commercialize some cultures into commodities. Commercialize the culture of wildlife tourism.

Sustainable wildlife tourism takes people's values into account. In other words, it recognizes a fair and ethical tourism industry through acceptable working conditions for everyone. Treatment of the living circumstances of locals with higher emphasis than tourist growth is the key element of sustainable wildlife tourism.

As a result, it completely inhibits inhumanity, including child labor, poor working conditions, human trafficking, and unjust salaries. Sustainable wildlife tourism focuses on improving inhabitants' standards of living by giving back to local economies through education scholarships, money-raising, investing in local initiatives, and employment.

Sustainable wildlife tourism encourages wildlife travelers to assist local people by purchasing local goods from farmers, dining at local cafes, and staying in local institutions. Wildlife visitors should purchase items from firms that support communities and also explore notorious locations instead of the areas that are commonly frequented by wildlife tourists.

Economic Sustainability

Economic sustainability is sometimes termed the economic pillar or economic health of protected areas or wildlife tourism attractions. In this context, it promotes quality of life for tourism locations to continue attracting wildlife visitors. It notably strives to recruit excellent and locally owned companies while supporting wildlife tourism activities.

Sustainable wildlife tourism guarantees robust and long-term economic operations by offering socio-economic advantages that are equitably allocated to all stakeholders. It provides secure employment and income. Wildlife tourism firms' current employment is one of the fundamental components of a sustainable economy. It plays a vital role in reducing poverty by providing social services, including assistance to local businesses in adopting sustainable principles.

Sustainability may allow local and small businesses to expand their formal and informal job opportunities. It maximizes local communities' economic independence by empowering and supporting local businesses. Sustainable wildlife tourism encourages travelers to purchase local services and goods as a way to support the activities of local industries and enterprises.

The purchase of souvenirs, food, and other locally produced items makes locals autonomous and inventive. Sustainable wildlife tourism also fosters fair collaborations between micro-local and international firms. As a method of assisting and empowering small tourist enterprises, multinational corporations give technical and financial assistance to local and small firms to boost their productivity and efficiency.

A solid, sustainable, and healthy economy relies on strong relationships with educational, non-profit, and global corporations. Moreover, sustainable wildlife tourism utilizes globally established and evaluated rules for training and certification. It alerts tourism stakeholders to the need for regular training and certification of their goods and services in accordance with sustainable tourism principles.

Sustainable wildlife tourism fosters the implementation of several national, regional, and worldwide accreditation systems. The programs foster corporate ethics and innovation while also rewarding the contributions of wildlife tourism actors to economic growth. Furthermore, they encourage the diversification of a wide range of wildlife tourism activities.

Wildlife watching, or game drives, has been a traditional or major activity in wildlife tourism. Sustainable wildlife tourism demands the identification and implementation of other sorts of tourist activities to optimize the tourism experience while also utilizing other wildlife resources correctly. Sustainable wildlife tourism reduces the enormous demand for limited local resources such as energy, food, and other raw materials.

The seasonal nature of wildlife tourism, when the destination is required to have up to 10 times more people in the high season than it is in the low season, may impose pressure on the demand and usage of local resources. Economic sustainability is all about the effective, efficient, and responsible exploitation of wildlife resources without violation of users' rights or the capacity of other stakeholders to utilize the resources.

Instead of concentrating on the rewards, the business entities design their harmless tactics for people and the environment. Sustainable wildlife tourism encourages stakeholders to strategically consider their spending and buying, keeping in mind that every dollar spent produces beneficial improvements to the community, protected areas, and the government.

It supports local investments by local individuals in supporting local communities and local companies. Sustainable tourism's role is to allow and empower local people to invest in visitor lodgings, facilities, cafes, attractions, and tour guides. Instead of soliciting, depending on, and supporting more international investors than local ones, sustainable wildlife tourism empowers and prioritizes local investors over foreign ones.

Moreover, it sensitizes wildlife visitors to remain in local lodging facilities and discourages overnight stays and expenditures in multinational hotel chains since it does not deliver important advantages to the local economy. The multinational hotels operate in enclave tourist models that barely regard the people.

Also, sustainable wildlife tourism encourages the use of local tour guides for interpretation and leads wildlife visitors to diverse wildlife hotspots. By doing so, wildlife visitors help the site's livelihoods and socio-economic development efforts.

This form of assistance, by providing many advantages to the community, may ensure the growth of wildlife tourism.

Strategies for Sustainable Wildlife Tourism

Strategies are the daily plans that the organization employs to execute the concept of sustainable wildlife tourism development. Destinations, protected areas, wildlife tour operators, and other wildlife tourism stakeholders must implement the stated principles of sustainable wildlife tourism. As strategies for the adoption of sustainable wildlife tourism, any location or wildlife stakeholders may have distinct plans for the adoption of principles of sustainable wildlife tourism.

Without practical, quantifiable, trustworthy, and realistic solutions, the principles of sustainable wildlife tourism are impossible to apply. Some of the generally proposed solutions for sustainable wildlife tourism development include tourist carrying capacity, tourism carrying capabilities, limits of acceptable change, management zoning, permits, and licenses, environmental quality standards, and visitor management systems.

Tourism Carrying Capacity

Tourism carrying capacity is the maximum number of wildlife visitors a place can sustain at one time without or with minimal social, ecological, economic, and cultural repercussions. It is the maximum number of visitors that can fit into a place at any given moment with the fewest negative repercussions.

According to the WTC, carrying capacity is the point at which a location or attraction begins to feel the negative effects of wildlife tourism. It is a time when social, economic, or environmental specialists start identifying the unfavorable repercussions of wildlife tourism at the destination or attraction. Any tourism activity above capacity becomes destructive and unsustainable.

The notion of tourist carrying capacity was derived from the domains of rangeland and wildlife management. According to the literature, the notion is dynamically complicated, and it takes substantial financial resources and varied technological expertise to accomplish it. If done correctly, it may help destinations create suitable rules and limit the consequences of wildlife tourism.

The WTC proposed four kinds of tourist carrying capacity: physical, social, economic, and biophysical carrying capacities, based on the types of effects. However, this book addresses three forms of tourist capacity, which are economic, environmental, and social carrying capacity.

Economic Carrying Capacity

This refers to the highest quantity of financial exchanges that a location or attraction can sustain without seriously causing environmental damage or depleting its natural resources. In other words, it is the greatest limit of economic expansion that an environment may sustain with minimal ecological consequences or loss of ecosystem services.

As far as sustainable wildlife tourism is concerned, the economic carrying capacity concept is vital as it balances economic expansion with environmental conservation. Some places, such as Bhutan, employ a high-value, low-impact sustainability strategy in which travelers pay a fee to control the number of tourists and help the local economy without causing severe environmental harm.

In summary, economic carrying capacity is the degree to which an economy can bear the negative effects of tourism activities. If tourist activities surpass economic carrying capacity, various unfavorable repercussions would occur to the local economy, such as loss of human labor in other sectors, economic leakage, unemployment, seasonal employment, overpopulation, and growing reliance on tourism, such as beggars.

When sites or attractions are particularly popular, they are likely to be overpopulated by wildlife visitors. Any overpopulated terrain is prone to environmental and infrastructural degradation. Additionally, such sites are prone to severe rivalry over local resources between visitors and inhabitants, which may lead to environmental devastation or overexploitation of natural resources.

According to the survey performed by MoneyTransfer, the most congested tourist destination in the world is Phuket, Thailand. According to the report, there are 118 visitors per local inhabitant. The overpopulation of Phuket Island highlights the need to adopt economic capacity. With such a ratio, the perceived detrimental environmental, social, and economic repercussions might be exceedingly high on the island if not regulated.

The seasonality of wildlife tourism may impact the location's socioeconomic activity and ecology. During the off-season, many locals become jobless; thus, it is conceivable for them to participate in environmentally harmful actions to preserve their lifestyles.

Also, an increase in employment during peak seasons may heighten environmental destruction as the need for environmental services such as wild fruits, vegetables, souvenirs, and bush meat grows to satisfy the rising demand from wildlife visitors. Harvesting and marketing natural resources generated from endangered species make wildlife tourism ecologically irresponsible.

Furthermore, economic carrying capacity helps prevent tourist economic leakage. It is one of the most common forms of wildlife tourism, especially in underdeveloped countries. Foreign wildlife tourism investors are typically involved in tourist leakage. It happens when the bulk of wildlife tourist income ends on foreign products and services.

Tourism leakage means that all earnings from wildlife tourism are exported without being reinvested in the local economy. Failure to reinvest the earned income in the destination or attraction leads to an uneven balance between economic development and environmental protection.

It is acceptable to say that these are some instances of traditional wildlife tourism's effects on the local economy. Such implications might vary depending on the location, attraction, and kind of wildlife tourist activity. Economy-carrying capacity is one of the essential components of preventing the detrimental consequences of wildlife tourism. The notion of economic carrying capacity may be extensively employed to regulate wildlife tourism for the advantage of the local economy without causing substantial ecological harm.

Social Carrying Capacity

Social carrying capacity is the maximum number of wildlife visitors that may visit the site or attraction without producing substantial adverse social consequences and diminishing the quality of tourists' happiness. It is a vital pillar or ingredient in sustainable wildlife tourism growth since it guarantees that wildlife tourism activities take place with social considerations in mind.

It is the extent to which the community can support a substantial number of visitors with minimal socially detrimental effects. When wildlife visitors exceed the existing social carrying capacity, the destination is likely to encounter unfavorable social repercussions, including child labor, drug abuse, and cultural commercialization.

As the major hosts of wildlife visitors and crucial participants in the protection of wildlife resources, wildlife tourism frequently sidelines and overlooks the social welfare of the people. Social carrying capacity is one way to ensure that wildlife tourism benefits residents without causing social problems.

Several variables may impact the social carrying capacity of the attraction or destination. The size of the host community, the number of wildlife tourists' visitations, cultural norms, and destination resilience are among the elements that may affect the social carrying capacity.

The larger the population size of the host community, the greater the number of wildlife tourists who may visit the area with little detrimental effects but also with a minor depreciation in the quality of tourist satisfaction. In the preceding paragraph, I listed Phuket in Thailand, according to a survey performed by MoneyTransfer, as the top congested location, where every resident engages with 118 visitors.

Though the published scientific estimation on the minimum agreed resident-tourist ratio is missing, the quoted one from Phuket seems boring. Such a large resident-tourist ratio may affect not only the environment but also the cultural features and satisfaction levels of wildlife visitors in the area.

The commodification of culture is one of the frequent and repeated influences since appealing to local culture is generally commodified for tourist purposes. As a result, the commodified culture of specific groups may lose its authenticity and identity as it becomes common and not desirable. By restricting tourists' visits, social and economic carrying capacity may safeguard sensitive cultural components and raise the quality of tourists' pleasure.

However, it is also vital to note that animal tourism areas do not equally suffer from visitor congestion. The resilience level of the destination may affect the degree of unfavorable social repercussions and deterioration in the quality of visitors' satisfaction at the destination. The attraction's durability may be determined at times.

Environmental Carrying Capacity

It is the maximum number of wildlife visitors the wildlife attraction can sustain without causing severe damage to the ecosystem or animal resources. It is the extent to which an ecosystem can handle a substantial number of visitors without or with minimal detrimental ecological repercussions.

As a result, it is also referred to as ecological carrying capacity. It is a vital idea in sustainable wildlife tourism since it warns wildlife tourism actors to balance all tourist activities with environmental conservation. According to the pillars, the number of wildlife visitors and amenities should not conflict with environmental quality and capability.

The concept is quite complex because it relates to ecological theories. Consequently, many wildlife-protected areas have limited the maximum number of wildlife visitors permitted to enter a particular management zone or complete protected areas at the same time.

For example, the government of Peru has restricted the number of visitors climbing the Inca Trail to 500 people each day. The 2010–2020 general management plan of Mkomazi National Park in Tanzania limited the maximum number of walking safaris to wilderness and high-use zones, where the maximum number of wildlife tourists allowed in walking safaris per group was eight and the maximum number of walking safaris per day was only four.

Several variables, including ecological indicators such as habituation, soil compaction, animal dispersal, and soil erosion, may influence the environmental carrying capacity. The limits on the number of hikers enforced by the Peruvian government originated from the geologists' concern that a large number may cause major soil erosion.

It is crucial to remember that exceeding environmental carrying capacity leads to major environmental concerns. These are the environmental challenges that we now encounter in traditional wildlife tourism. Such challenges include water and air pollution, the extinction of desirable wildlife species, the alteration of animal habits, and the depletion of wildlife habitats.

It is important to note that the environmental or ecological carrying capacity of the protected areas relies on numerous criteria, including the area, wildlife tourism activities, size, scale, location, and possible unfavorable consequences. In the preceding paragraph, Mkomazi National Park established the carrying capacity for certain activities in a given zone, wilderness, and high-usage zones.

Regardless of the zones, certain protected areas may establish environmental carrying capacity for the whole protected area for more than one day. For example, in Norway, Dovrefjell-Sunndalsfjella National Park has an environmental carrying capacity of 12 visits per square kilometer per year. In Germany, the Biosphere Reserve in southeast Rügen has an environmental carrying capacity of 8000 wildlife visitors per square kilometer per year.

Weaknesses of Tourism Carrying Capacity

It is simple, both theoretically and practically. It demands significant study, which takes knowledge and considerable time. If it requires investigation, it is difficult to assess a carrying capacity since it also relies on other elements, including visitor habits, economics, and political and cultural considerations. It is expensive to adopt in terms of financial and time concerns.

Limits of Acceptable Use

The Limit of Acceptable Use or Change (LAC) is a strategy for the implementation of sustainable wildlife tourism that relies on defining quantifiable anthropogenic visible and invisible social, economic, and ecological changes induced by wildlife tourist activities at the destination.

In certain wildlife tourist attractions, LAC is a participatory planning principle derived from ecological research that uses baseline and ideal circumstances to identify the harmful consequences of wildlife tourism.

It is a thorough reformulation of the notion of tourism carrying capacity, but in this context, the major area of attention is the preservation of desirable conditions rather than the quantity of visitors that a place can accept.

It entails thorough research or inquiry to determine all the desired and necessary social, environmental, and economic characteristics of the destination or attraction in order to identify different degrees of wildlife tourism utilization and destruction.

After determining the ideal circumstances, tourism experts identified desirable social, environmental, and economic factors as indicators of tourists' usage at the locations. Any departure from the planned circumstances may be extensively exploited by specialists as a warning or signal of strange things occurring from wildlife tourism at the site.

Some of the indicators frequently used in the Limit of Acceptable Use include changes in the structure and features of vegetation, strange behaviors of wild animals, children misbehaving, changes in the living standards at the destination, black marketing, excess social crimes, and excess conflicts between local communities and tourism enterprises, among others.

Apart from those described above, other indications for identifying anomalous effects of wildlife tourism exist. Adoption of such metrics differs across locations due to several variables, including the destination's resilience, geography, and degree of economic growth. Tourist specialists may use either a few or many established markers to estimate the level at which wildlife tourism affects the location.

Wildlife ecologists and foresters are competent in utilizing the circumstances of vegetation to identify the site use and features of resource utilization in wildlife tourism. Changes in vegetation color, shape, position, length, and girth may reflect wildlife attractions' ecosystems or vegetation's health. Wildlife tourism has an impact on the physiological and behavioral characteristics of wild plants and vegetation, such as pollination and blooming.

Conventional wildlife tourism activities, such as off-road driving, the introduction of foreign invasive species, wildfire, casual littering, and deforestation, may harm the anatomy of wild plants and the vegetation structure. Careful monitoring of wildlife tourism activities offers wildlife tourism professionals insightful indications of the present state of the vegetation and relevant mitigation actions to be taken.

Also, wildlife tourism experts set or change the extent of site use after noting the emergence of strange or abnormal behaviors from wild animals, such as wild animals begging for food, wild animals queuing along the tourist roads, loss of fear of humans, and wild animals feeding on strange food items. Wild animals' begging behaviors may result from overfeeding by wildlife visitors, off-road driving, poaching pressure, and inadequate waste management by tourist facilities.

Moreover, wildlife tourism specialists need to govern the use of particular attractions or destinations by either expanding or lowering the amount of their utilization by wildlife visitors. Strange actions from youngsters and adolescents are some of the main qualities that demand the relevant authorities determine the limit of destination or attraction utilization by wildlife visitors. The clothing style, engagement in drug misuse, prostitution, students quitting schools, augmentation of street children, and employment of minors in tourism facilities.

Furthermore, changes in living conditions at the destination might cause the setting up of restrictions on destination usage, as it reflects the incapacity of the destination to deal with the rising need for products and services from wildlife visitors. Insufficient supply of products and services at the destination level, owing to the increasing number and spending of wildlife visitors, may lead to a severe shortage of vital commodities, rendering people unable to afford to acquire critical commodities.

The scarcity of goods leads to an increase in commodity prices, resulting in higher living standards at wildlife tourism attractions. As a result, residents struggle to cope with rising living standards and forsake their traditional socio-economic activities. With such indications, destination management organizations establish a suitable limit on destination usage to rescue the locals from the unforeseen, unfavorable socio-economic repercussions of wildlife tourism.

However, determining the limit of use for the destination or location is not as straightforward as it seems, since it needs multiple levels of awareness, engagement, administrative choices, expertise, and research. According to their requirements, the location or attraction must be satisfactory. Among other things, the destination should identify the types, baselines, and acceptable conditions of resources against which the circumstances will be compared.

The establishment of the existing socio-cultural and environmental resource conditions before the implementation of LAC is vital, as it may provide thoughts on the advancement of resource utilization and circumstances as benchmarks to compare with. Understanding the resources' circumstances may require long-term research and working experience with the attraction and protected region.

The state or circumstances of the attractions, such as the species richness and number of plants and animals, the cultural conditions of the populations, and the economic status of the destinations, are crucial benchmarking criteria. The established baseline conditions may be received from numerous sources, including destination management organizations, governmental and non-governmental organizations, and research institutes.

Also, establishing acceptable or benchmarking conditions for the resources establishes their use restrictions. Acceptable circumstances, also known as benchmarking conditions, are crucial in comprehending the linkages between the existing and prospective resources' conditions. In other words, desired or benchmarking situations may be expressed through visible or unseen aspects such as colors, fragrances, and resource behaviors.

In such a situation, any alteration in the color, smell, or behavior of resources may be considered to exceed the desired or acceptable conditions of the resources arising from wildlife tourist activities. As a consequence of wildlife tourism, tourist managers may use the specified desirable conditions to determine the level of resource utilization at the destination or protected areas.

The limit of permissible usage or change does not stop with the establishment of baseline and desired conditions alone; it continues further to the identification and execution of management activities for the accomplishment of the sustainability aim of wildlife tourism. Identification of proper policy and legal action mechanisms for the execution of LAU may prevent wildlife tourism from surpassing the desired or benchmarking conditions.

Furthermore, monitoring and assessing the management measures for implementing the limit on permitted usage is critical. The monitoring aids in determining whether the wildlife tourism stakeholders perform the correct things using the appropriate methods. In other words, it ensures that the stakeholders apply the LAU concept according to the agreed plans.

Management Zoning

Management zoning involves segmenting the destination or protected area into various parts depending on the degree of usage and management intensities. It is a typical participatory destination and protected design method that aims to link forms of development or wildlife tourist activities to distinct regions, such as low-usage zones, wildness zones, entertainment zones, and cultural zones. Wildlife tourism stakeholders decide on the number and types of management zones that protected areas, attractions, or destinations may have.

Wildness Zone

It is a management zone dominated by wildness; such regions are rarely exposed to wildlife tourist activities. The existence of highly endangered species, limited knowledge of species variety and abundance, and the absence of suitable wildlife tourism facilities and infrastructure inside wilderness zones are some of the issues that impede wilderness zones from attracting wildlife tourism activities. In most instances, destination management organizations and protected areas scarcely entertain the development of designated wildlife tourism routes, hotels, and other amenities in the wilderness zones.

Entertainment Zone

The entertainment zone is one of the destination's most used management zones and may promote different levels, types of activities, amenities, and infrastructure supporting wildlife tourist activities at the destination. The zone encourages unusual and outdoor activities such as treasure hunting, airborne and hydro adventures, and diving tourism, among others. However, it is reasonable to state that, aside from being the widely used management zone, sustainable wildlife tourism constantly respects and respects the carrying capacity.

Cultural Management Zone

The cultural management zone is a frequent kind of zone, found in both protected and unprotected landscapes. It is the location where cultural activities take place and cultural resources are safeguarded. In this zone, only cultural activities are permitted as a means of improving and honoring traditional knowledge.

Tourism Permits and Licenses

Sustainable wildlife tourism advocates the use of permits and licenses for controlling diverse tourist activities at the destination and attraction. Such permissions are necessary to conduct different tourist enterprises based on location, season, and attraction. Wildlife tour providers need approval to transport visitors to wildlife sites as defined by legislation.

Depending on the kind of attraction, multiple authorities are responsible for issuing operating licenses to tour operators. Public agencies are responsible for issuing tour operating permits at public wildlife attractions, whereas private organizations grant permission to operators.

The same applies to wildlife tourists; proprietors of wildlife attractions may charge wildlife visitors a specific amount of money to regulate the number of tourists at the attractions. Depending on the attraction's sensitivity and fragility, the owner may raise the entry cost to preserve culturally or ecologically sensitive regions.

Similarly, the destination or attraction administrators may charge tourist equipment or facilities a specific amount of money to preserve the environment. In many protected areas, charging variable admission fees for automobiles based on the vehicle's weight and tailpipe carbon dioxide emissions is typical.

However, weight-based billing is the most frequent technique at many wildlife attractions since heavier tourist equipment is charged more than lighter ones. This encourages the use of light vehicles, boats, and planes in a fragile environment while also maintaining the fragile ecosystem.

In brief, the utilization of permits and licenses is crucial at the destination and attraction levels since the operators must get permission and licenses in advance to run lodging, transport, meals, and drinks at the destination. Such permits and licenses are subject to rules and compliance requirements.

Additionally, permits and licenses generate income for private, communal, and government entities because each operator and visitor pays money to obtain permission. The money gained may be used to finance conservation projects at animal attractions, assist local communities, and upgrade tourist infrastructure and superstructure.

Setting of Environmental Standards

Sustainable wildlife tourism demands a baseline of environmental criteria for wildlife tourism providers and sites to conform to. Also, the environmental criteria serve as guides for the implementation of wildlife tourist activities within and beyond the protected regions. All wildlife tourism activities must be established and administered under environmental quality requirements.

Some of the quality environmental standards are the maximum height and size of buildings in ecologically sensitive areas; the color of the buildings should be elaborated to blind the environment; restrictions on the use of construction materials by the enterprise; the location and placement of buildings at the destination; and the conduct of EIA before and after instituting any tourism enterprise.

Other quality environmental standards include carrying out environmental audits, promoting environmental offsetting projects for tourists, such as afforestation for carbon dioxide, and using ecologically friendly materials, including local construction materials, to preserve existing vegetation when constructing facilities and infrastructure for wildlife tourism.

Setting the maximum height and size of the tourist facilities is one of the quality environmental norms supported in ecologically sensitive zones. Both excessive downward and upward size or length of tourist amenities and infrastructure may interfere with ecological processes in protected landscapes, and the deep development of tourist facilities in terrestrial and aquatic environments may interfere with the hydrological system.

It is crucial to realize that the protected zones are intended to preserve all resources happening below, on, and above the ground. With that knowledge, the destination should define the maximum and minimum size of the tourist amenities and infrastructure, without or with minimal consequences. Additionally, the maximum distance between tourist facilities and water bodies should be defined to protect the aquatic environment from unsustainable tourism activities.

Moreover, there should be quality environmental requirements on the hue of tourists facilities; sustainable wildlife tourism supports the colors that blind the environment. To blind the environment, the hues of the structures should be elaborate. The laws must state specifically the needed and prohibited colors for lodging facilities, vehicles, balloons, and aircraft to operate in wildlife attractions.

Visitor Management System

Sustainable wildlife tourism promotes wildlife tourist locations to track the utilization of the site by collecting information, records, and site usage. Constant surveillance and monitoring of site utilization may give a hint to destination administrators about devising suitable plans for mitigation of any predicted harmful effects.

To avoid excessive visitation, as certain attractions or destinations are more frequently visited by wildlife tourists than others, destination management organizations should encourage the tourists to use other areas by either lowering the entrance fee to other areas or improving tourist infrastructure and facilities to attract more tourists.

Construction of state-of-the-art tourism infrastructure, diversification of wildlife tourist activities, and improvements to communications and transport services may encourage wildlife visitors to explore places that are infrequently frequented by wildlife tourists.

It balances wildlife tourism activities and also avoids geographical and temporal polarization, the key features of conventional or mass visitors. Also, wildlife tourism prohibits any activity, practice, or equipment that destroys nature, culture, and the economy.

However, identifying activities and equipment that harm the destinations is one thing, and minimizing the consequences is another. In such circumstances, the deployment of multiple measures, including studies and frequent monitoring, is necessary for the identification of related harmful consequences. Formulation and execution of solutions to reduce the observed consequences are equally crucial.

One of the solutions is applying variable admission fees to various tourism equipment and activities depending on their degrees of harmful consequences for the environment, society, and economy. For instance, the admission cost and operating tourism permits should greatly differ between light and heavy-weight tourists' trucks, aircraft, and boats. The admission fees may be levied differently depending on their weight, loudness, and tailpipe gas emissions.

To deter and reduce the levels of damage, the more damage, the higher the admission and use fees paid to the activity and equipment. Also, the application of appropriate taxes, fines, and penalties for the building of tourism amenities that may harm the environment, communities, and local economies is another possibility.

Discourage the construction of tourist facilities in ecologically sensitive areas and culturally sensitive places, which include areas near schools, worshipping places, sources of rivers, breeding areas, migratory corridors, and dispensaries, by either subjecting the construction of tourist facilities to EIA before and after construction or penalizing any tourist facility constructed without following the appropriate procedure.

Sustainable wildlife tourism comprises educating wildlife tourists about wilderness ethics as a method of familiarising people with nature protection. It is unethical for unscrupulous wildlife tour operators or guides to overlook informing wildlife visitors before and throughout their journeys about the ethics and rules of wildness. Constant and timely instruction on wilderness ethics supports the compliance of wildlife visitors with the legislation.

Conflict Resolutions and Trade-Offs

Conflict resolution and trade-offs are the toughest areas of wildlife conservation and wildlife tourism. Conflicts between humans and animals rank second in wildlife conservation. The conservation cost that local populations pay for animal protection is undoubtedly priceless. For wildlife tourism to survive, it must overcome all obstacles. Tradeoffs arise when the fulfillment of one goal contradicts a significant other.

In some instances, the expansion and development of wildlife tourism may enhance wildlife trafficking, poaching, and environmental harm. Alternatively, growth in the animal population may lead to human–wildlife conflicts. Wildlife tourism has to have a clear framework for dispute resolution and compromises. Wildlife tourism has both detrimental and positive elements. Wildlife tourism, if intelligently managed, would benefit ecology, society, and the economy by generating money and supporting livelihoods and cultural values.

The detrimental side of wildlife tourism is unsettling, as it substantially inflicts adverse ecological consequences on protected areas, intensifies human–animal conflicts, and generates local inflation. Conventional wildlife tourism inflicts unending costs on wildlife stakeholders; animals, as the primary resource for wildlife tourist development, fall into endless confrontations with humans. Local communities are the major victims and losers of the conflict; the loss of property and human lives to animals is the main expression of human–wildlife conflicts.

When wildlife diversity and abundance increase, conflicts become disastrous. To boost positive outcomes and decrease negative consequences, conflict resolution and tradeoffs are critical factors of sustainable wildlife tourism for balancing the environmental, community, and development goals of key stakeholders in wildlife tourism. Some examples of tradeoffs and conflict resolution tactics include community participation, conservation costs, and benefit sharing.

Sustainable wildlife tourism tries to balance the requirements of primary wildlife tourism stakeholders by engaging them actively in the conservation and usage of animal resources.

One means of sharing cost benefits from wildlife tourism is the decentralization of wildlife conservation to communities. In this environment, the local people manage animal resources in their areas without or with minor interference from the government. It creates a sense of ownership and trust for local people because they regard animals as property.

Similarly, it promotes the tolerance and resilience of local people against animal destruction and annoyances. To raise earnings from wildlife tourism, local people may strengthen their efforts towards wildlife conservation by providing hands to conservation agencies in the ongoing struggle against wildlife crimes.

It is vital to underline that sustainable wildlife tourism also encourages the wildlife tourism stakeholders to contribute to the socio-economic development of the communities near the wildlife attractions as a way of appreciating their support for wildlife conservation.

Is Sustainable Wildlife Tourism Simple?

It is the question that wildlife tourism stakeholders ask routinely. It interrogates whether the viability and ease of adoption and implementation of all pillars of sustainable wildlife tourism are straightforward. It is easy to talk about and prepare for sustainable wildlife tourism, but it is always difficult to implement. As a result, some organizations or institutions embrace just one, preferably the social pillar, while neglecting the other pillars.

According to the literature, many organizations, enterprises, and destinations embrace sustainable tourism conceptually, not practically. Many newly rising locations talk about sustainability, but they still employ the same old habits with awfully harmful repercussions.

However, such groups have valid reasons for not adopting the concepts of sustainable wildlife tourism growth. The literature lists economic, institutional, and business issues as some of the issues that restrict various organisations from adopting and applying the concepts of sustainability.

Most wildlife tourism destinations focus on improving the locals' economic well-being through economic development and job creation. Consequently, economic gains rate as a greater national priority than environmental protection and socio-cultural warfare.

Many locations find it difficult to reject promising strategic initiatives at the expense of environmental protection or social elements. It is clear from various wildlife tourism areas that commercial ventures devastatingly destroy the delicate ecosystem and endanger the locals' social warfare.

Sustainable wildlife tourism demands that all responsible institutions, organizations, or corporations adopt all the pillars or principles of sustainability but not one or two and neglect the rest. Failure to achieve either of the pillars of sustainability in wildlife tourism will not make it viable any longer.

Incidents arise when certain organizations or enterprises select just one of the pillars of sustainability for political or financial gain. Such organizations frequently choose social pillars to earn society's trust by providing a few social services to communities while destroying the environment.

It is also a chance for certain commercial organizations to optimize their business advantages by utilizing the sustainability idea as a marketing technique. It is possible to locate various tourism facilities created with environmentally friendly materials, yet the same facilities are full of half-naked visitors or entertain child labor.

Also, the displacement of residents' land for the construction of tourism ecolodges and other amenities for wildlife visitors causes the villagers to lose their ancestral houses and traditional locations such as worship grounds. It is a significant infringement on the social pillar of sustainable wildlife tourism. It disrespects the customs of the locals.

It is possible to find a wildlife tourism firm, such as a hotel or tour-running company, that adheres well to social and environmental pillars but fails to meet economic sustainability standards. Such firms frequently employ local guides, use recyclable building materials, serve organic cuisine, and are socially responsible.

However, such firms are 100% owned by overseas corporations, which refund all generated revenues to the owner's overseas account. Moreover, the wildlife visitors have minimal connections with local populations, and the supplies for amenities, including food, groceries, and toiletries, are imported from overseas. In this scenario, the facility fails to fulfill the economic concept of sustainability.

Finally, it is challenging to execute all of the pillars of sustainable wildlife tourism in terms of self-commitment and realization. Most wildlife tourism firms regard the execution of all pillars as a hardship. The apparent factors include a lack of proper knowledge and skills and a desire to execute the pillars. It is a significant infringement on the social pillar of sustainable wildlife tourism. It disrespects the customs of the locals.

Chapter 6
Planning for Tourism Sustainability

Abstract Also, the environment highlights the fact that the wildebeest migration relies on both nations for its existence. That is to say, the use of wildebeest resources depends on both local planning at the scale of Masai Mara National Reserve and Serengeti National Park, as well as national planning at the level of the Kenya Wildlife Planning Corporate Plan and Tanzania National Park Corporate Plan. This chapter underlines the significance of participatory planning for sustainability at global, regional, national, destination, and attraction levels in the context of responsible tourism. In the framework of responsible wildlife tourism, it makes clear the mutually beneficial relationship that exists between sustainability and protected areas. It also underlines the necessity of laws, monitoring, and assessment of wildlife tourist operations in line with the philosophy of sustainable development. The consumption of resources without scientific and participatory planning causes damage to the economy and wildlife ecosystem. Successful and feasible plans usually depend on appropriate integration across various national, regional, international, and ethnical scales, which is important for managing national and transboundary wildlife resources. Without proper and devoted implementation techniques, sustainable wildlife tourism may develop in a haphazard way. As a result, it is mandatory for certain governments for each protected area to have a management plan that technically and legally guides resource utilization and conservation at the attraction and destination levels. Most of the management plans are developed in the participatory approaches by incorporating many key wildlife stakeholders. Such plans provide the basis for managerial decisions and future directions of the protected areas. Furthermore, the management plans may have management zones outlining and limiting the execution of specific operations based on resource sensitivity and availability. Planning is crucial for reducing negative consequences on sensitive resources and endangered species. The systematic planning process in wildlife tourism occurs as a lengthy and cyclic process, not as an event. It involves every essential wildlife tourism stakeholder and follows all the necessary but scientific participatory planning processes. A proper and participatory planned sustainable wildlife tourism provides accurate information to stakeholders and maximizes the quality, accessibility, safety, and security of tourism products and services at the destination. Consideration of the expectation, safety, motivation, interest, and

A. A. Mamboleo, *Sustainable Wildlife Tourism*,
https://doi.org/10.1007/978-3-031-97795-4_6

satisfaction of wildlife tourists is one of the outcomes of a well-planned sustainable wildlife tourism.

Keywords Sustainable development · Ecological integrity · Community empowerment · Economic resilience · Environmental impact assessment · Stakeholder collaboration

Participatory planning is one of the sustainable techniques for resource usage. Without scientific and participatory planning, resource utilization may occur in an unplanned or unsystematic manner. Unplanned tourist activities may be damaging not only to the environment but also to local people's livelihoods.

Even the planning should be part of the local, national, and even regional business strategies. The integration of wildlife tourism operations at multiple geographical scales and land uses may boost efficiency and decrease the detrimental consequences of wildlife tourism. Because sustainable wildlife tourism must adhere to sustainability principles, planning for tourist operations is necessary.

If the protected areas are not scientifically planned, their management might be unproductive and inefficient. When planning for tourist activities, different levels and sizes of planning are required. At the local scale, wildlife tourism should be part of protected areas, where specifications on the kind, zones, and limits of resource utilization for sustainable wildlife tourism inside a particular protected area are highlighted.

Moreover, the protected areas may be part of regional networks of protected areas or regional ecosystems; this is highly practical for many protected places located in transboundary ecosystems. To clarify, a transboundary ecosystem includes the political borders of more than one nation or state.

The Serengeti-Masai-Mara ecosystem encompasses Tanzania and Kenya; the Mkomazi-Tsavo ecosystem covers Tanzania and Kenya; and the Selous Niassa ecosystem spans Tanzania and Mozambique. Those are a few instances of transboundary ecosystems in Africa. In a transborder ecosystem, wildlife resources evolve into regional, not national, resources, and in many circumstances, the resources can rarely live without crossing the national frontiers.

Moreover, and significantly, the existence of animal movement relies on regional-level planning. This level features mostly Tanzania and Kenya. The use of wildlife resources in such places requires the wisdom and concentrated efforts of concerned nations since migratory animal species cross borders without understanding the political authority of the countries.

After a thorough discussion on the importance of wildlife tourism planning, let us simplify the process by explaining how it may occur at the local level in a protected area. In certain nations, it is a law for every protected area to have an active management plan before formal gazettement. A broad management plan lays out the management methods and objectives for every resource and zone.

It is a framework for decision-making about resource management in the protected area over a defined period. Because it is comprehensive and covers every resource, the recorded plan for the administration of a specific protected area is commonly referred to as a general management plan, or GMP. It is frequently more or less detailed, depending on the purpose for which it is used and the legal criteria to be met.

The Essence of Planning for Sustainability

Wildlife tourism activities must be part of the wildlife tourism strategy, although not all destinations and attractions are prepared for the sustainability of wildlife tourism. It is necessary for all sites with active wildlife tourism activities to likewise have an active wildlife tourism strategy as a way of directing their operations. But especially before implementing the sustainability idea, tourist planning is vital, independent of its elements at the destination.

Without suitable wildlife tourism planning, the preservation, development, and use of wildlife resources would proceed haphazardly, frequently in response to political or tourist demands, with little thought as to the implications for the future or sustainable development.

As a result, several governments mandate that every destination or attraction have an active and adequate wildlife tourism management strategy. Sustainable tourism planning is a result of the participatory planning process, which involves recording the management strategy, the choices made, the justification for these decisions, and the direction for future management of varied wildlife tourism activities.

In such a scenario, it is vital to have a tourist management plan, as it enhances the administration of protected areas. Consequently, wildlife tourism managers may connect the wildlife tourism strategy to "the constitution of the destination or protected area." As all management decisions regarding wildlife tourist activities rely on a thorough knowledge of the location,

As planning for protected areas entails a thorough study and resource inventory in the protected areas, it provides the planning team with adequate and current knowledge about the protected region. In such circumstances, the destination or attraction's administration became methodical and open.

Also, the wildlife tourism management plan provides clear rules for managing wildlife resources as it describes the type of activity, time of implementation, site of implementation, and accountable person. With such conditions, it becomes easier for managers to understand the sequence of operations, monitor the usage of protected areas, and monitor all activities.

As a result, in many protected places, wildlife tourist activities may occur only in certain zones. Some destinations have been broken into tiny pieces or zones called management zones, subdivisions based on similarities and differences in resource uniqueness, richness, resilience, and sensitivity of the region.

The wildlife tourism strategy ensures management consistency. Without a proper management strategy, there is a constant repetition of chores. It gives a chance to identify, monitor, and assess the percentage of completion for each job throughout the management plan.

That is to say, any unfilled duty may be passed to the following wildlife plan: This eliminates task duplication because the cleared tasks will not be included in the next planning phase. I will also illustrate the continuity of administration by offering specific instances of protected areas as tourist attractions that fall under the umbrella of one conservation organization.

For instance, all national parks in Tanzania are under Tanzania National Parks (TANAPA); in Kenya, the protected areas are under Kenya Wildlife Services (KWS); and in the Republic of South Africa, they are under SAN-PARKS. Sometimes, conservation organizations relocate their workers across parks as a method of boosting efficacy and efficiency in the parks.

Following the transfer, the new staff normally do not know where to start in the new workplace until they come across a certain document. The general management plan may aid the personnel in the new park since everything about the park is already in the general management plan.

Moreover, the wildlife tourism strategy helps determine management efficacy and efficiency. In other words, efficacy and efficiency reveal that protected area administrators are doing the right thing. If so, are they doing it in the right way? The management plan is the proper technique to understand whether the management of the protected areas is on the right track. It consequently examines the efficacy and efficiency of each resource in the protected regions.

Furthermore, the strategy enhances the utilization of financial and personnel resources. The protected areas include animal and non-wildlife resources. In this context, non-wildlife resources, notably human and financial resources, are of significant importance. Since the management plan defines and records all actions for its whole lifetime, for more than five years, such documentation also specifies the amount and cost of required resources for the execution of the plan. This eliminates wasteful resource usage because everything should go according to plan.

Planning for Wildlife Attractions and Destinations

The preceding parts of this chapter have carefully explored the significance and planning of sustainable wildlife tourism development. In the core, wildlife tourism happens inside the wildlife tourist destination but not everywhere in the destination. It occurs in wildlife areas, although most of them are not primarily intended for recreational purposes.

Uncontrolled wildlife tourism activities are harmful to the integrity of protected areas and the future of endangered and endemic animal resources, and they are also opposed to the principles of sustainability. Interestingly, most categories of excellent resource values inside many protected areas are the major assets for wildlife tourism growth and diversification.

Clarifications are critical for everyone to understand the links between protected areas and sustainable wildlife tourism planning. Arguably, the links between the two occur on two sides: wildlife tourism as the result of wildlife resources and wildlife tourism planning as the strategy for sustainable wildlife tourism.

All two entities are vital for the development of wildlife tourism, as they may boost the quantity and presence of wildlife resources. Without consideration of the aforesaid elements, wildlife resources may be used unwisely, consequently compromising their sustainability and production. And without protected areas, wildlife tourism would be unsuitably implemented.

As the product or consumer of wildlife resources, regulation and monitoring of wildlife tourism is vital for all wildlife tourism activities to match the availability, abundance, fragility, carrying capacity, and resilience of animal resources. As a result, within the same protected area, wildlife tourists may be permitted to access specific areas or animal resources while being restricted to certain parts.

It is because granting wildlife tourists access to ecological and culturally sensitive locations where inadequate knowledge of their prospective impacts is either yet unknown or understood may be hazardous. Conservationists, on the other hand, include wildlife tourism in the protected area's planning process to supervise and safeguard its detrimental repercussions against endangered species or sensitive Exceptional Resource Values.

During planning, the conservationists may designate specific forms of wildlife tourist activities to a specified zone, time, and limit. It is impossible, as well as immoral, to allot or combine wildlife watching and trophy hunting in the same region. I suppose you comprehend the likely awful disasters. Since all of these activities occur in the same protected region, it is critical to properly organize them for smooth operations as well as prevent any negative effects.

Consideration of wildlife tourism in protected areas' planning is necessary since sustainable wildlife tourism has to be properly managed and always tends to prevent any harmful social, economic, and environmental implications. Wildlife tourism's sustainability is unachievable without considering the three pillars of sustainable development.

In short, our discussion about wildlife tourism must focus on sustainable development ideas, which is one of the techniques for the implementation of sustainable tourism development. As indicated by the Brundtland report, participatory planning is the inclusion of diverse stakeholders in the planning process.

That is to say, while planning a protected area, it is important to include the actual involvement of different key stakeholders, including wildlife and tourist stakeholders. The stakeholders have direct and indirect impacts on the protected areas. But in this book, let us address the planning for sustainable wildlife tourism in the context of protected area planning.

The planning for sustainable wildlife tourism is inherently participatory, and the input of many important stakeholders impacted by the administration of the protected area is required. In other words, the engagement of important stakeholders should take place as early as possible and continue throughout the planning phase of wildlife tourism.

The plan is considerably more likely to be executed if the impacted audiences are engaged in its formulation and have a feeling of shared ownership. In this sense, engagement also includes the integration of ideas from both formal and informal institutions. Local organizations, like traditional groups, are frequently ignored in these procedures.

This book discussed the wildlife tourism planning process as a simple topic, explicitly citing Mahale Mountains National Park as the guide for the debates. The protected area is situated in Tanzania's western tourist zone. At the protected area level, planning for sustainable wildlife tourism can be effectively implemented.

In most circumstances, the special committee may lead or manage the whole planning process. The committee may be composed of either professional advisors or experienced personnel from the protected area. The terms of reference generally direct the planning committee on which route to proceed in the planning process.

The phrases function as goals of the planning process, although they are often problematic when it comes to determining the management objectives of the protected areas. Are the management goals determined before or after doing research or inventory? Both techniques are viable; however, study and inventory are highly suggested since they may offer a clear image, maybe for additional protected areas, of the kinds and abundance of all animal resources in the protected areas.

Among other activities, the planning team initially identifies important stakeholders to aid in the planning, execution, monitoring, and assessment of the management plan. The stakeholder analysis should be carefully conducted to identify various stakeholders with diverse interests in the protected region.

Some stakeholders in wildlife tourism are communities around the protected area; domestic and international tourists; politicians; local and international non-governmental organizations; community-based organizations; the protected area's staff; the government; research institutions; and education institutions.

It is also significant to mention that planning for sustainable wildlife tourism is a continuous process with three basic aspects: preparation of a management plan, implementation of the plan, and monitoring and revision of the plan. However, the actual process of planning may be divided into numerous discrete parts.

These phases simplify the approach to performing the planning process; in this instance, the following are systematically stated below:

Step one, or the first step, is commonly referred to as the pee-planning phase, which is one of the most important processes in the planning process. This stage describes what the process will accomplish, how it will be carried out, scheduling concerns, and who is to be engaged. These decisions must be made at the highest possible management level and are critical to starting the planning process on the right foot.

Stakeholder identification is critical because accurate data and inventory are key practices for planning and maintaining the protected area. The planning team may utilize primary and secondary data to determine the resources inside the protected zones. These data can be gathered from both primary and secondary sources.

These statistics provide all of the facts about the particular protected region that are significant in estimating its importance. The importance of the protected region hinges on its extraordinary resource values. It is vital to recognize that the exceptional resource value is unique and located in the protected region.

After inventorying all exceptional resources, the established data may be used to calculate the exceptional resource values. Because we have chosen Mahale Mountain National Park as our case study, the park's general management plan separates the park's exceptional resource values into natural resource values, scenic resource values, social resource values, and cultural resource values.

In this situation, the exceptional resource values are unique and desirable resources discovered in the park. How did we know whether they existed? Remember how much research and inventory were meticulously completed earlier? Who did the inventory and research? It depends on the type of research and inventory, and here is where the notion of stakeholders' involvement comes into play.

It is not feasible for either the protected area or the relevant conservation agency to perform all or every study; instead, research institutes like universities or specialized research institutions may aid with research. One more question: how do exceptional resource values link with sustainable wildlife tourism?

In brief, exceptional resource values are the magnets of wildlife visitors in any protected area; they are living and non-living things located inside the protected area, not only for attracting tourists but also for conservation for the following generations to access. I believe we still recall what we covered in the opening chapter regarding the importance of protected areas, a crucial aspect when thinking about creating protected areas.

The importance lies in the combination of numerous exceptional resource values discovered in the safeguarded area. That is to say, we need to identify all exceptional resource values before assessing the importance of the protected region. After a brief explanation of the values, let us study and debate the classification of exceptional resource values as defined in the general management plan of Mahale Mountains National Park, which are natural, scenic, social, and cultural resources.

Natural Exceptional Resource Values

The park chose to classify all detected living organisms in the park into natural resource values. However, it is crucial to mention that there are no unique criteria for classifying exceptional resource values in the protected region. Such categorizations may differ within protected areas—even those overseen by the same conservation body.

We have seen that wild plants and animals are under the natural resources category in Mahale Mountains National Park; however, other protected places might label or group the same resources under Biological Exceptional Resource Values. After a thorough analysis, the data indicated that Mahale Mountains National Park has Tanzania's largest population of protected chimpanzees.

Chimpanzees inhabit 12 African countries. The eastern subspecies of chimpanzee found at Mahale Mountains National Park is classified by the IUCN as endangered. It is estimated that around 700 out of 1800 chimpanzees in Tanzania inhabit Mahale Mountains National Park, which is around twice the size of any other chimpanzee sub-population in Tanzania.

It is an outstanding resource, particularly at local, national, and worldwide levels. The chimpanzee's uniqueness as a natural feature necessitates specific conservation activities as a means of enhancing its values and existence, but it is also a natural magnet for wildlife tourists.

Scenic Resource Values

These are a mix of geographical elements that function as animal habitats and also provide recreational benefits for visitors in the Mahale Mountains National Park. The beaches near Lake Tanganyika are one of the typical scenic treasures. The beaches offer a beautiful recreational environment and sites where wildlife visitors may conduct various forms of tourism activities.

The northern, and most accessible, area of the park contains a number of pristine beaches. The unspoiled nature of these beaches, set between the steep rise of the Mahale Mountains behind and the majestic expanse of Lake Tanganyika in front, provides an idyllic location for park visitors to enjoy sunbathing, swimming, snorkeling, and picnicking.

Lake Tanganyika is another exceptional scenic resource value in the Mahale Mountains National Park. It is a unique and vital natural resource because it is home to many rare and endangered freshwater species around the world, including Lake Tanganyika Electric Catfish. It offers an ideal setting to practice and study different sorts of wildlife tourism. According to the management plan:

> Set at about 780 m above sea level, Lake Tanganyika is the longest in the world (720 km) and Tourism at Mahale Mountains National Park has been steadily increasing over the last few years, and this trend looks set to continue bringing benefits to both the national and local economies.

Such descriptions make the lake distinctive and special at the local, national, and worldwide levels. It is vital to recall that the lake contains a transboundary aquatic habitat constituting many nations, including the Democratic Republic of the Congo, Tanzania, Zambia, and Burundi. Refer to our previous in-depth discussion on transboundary habitats and ecosystems, as well as the importance of protecting them.

Socio-Economic Resource Values

The importance of the protected region may also extend to social and economic issues. Our debate on the sustainability of protected areas emphasized social fairness as one of the primary concerns of sustainability.

During the planning study and inventory, also evaluate the distinctive social value the protected places may offer to local and international groups. The availability of fish-spawning places is one of Mahale Mountains National Park's exceptional social values.

> Mahale Mountains National Park protects important fish breeding grounds and fish stocks that are vital to the continued health of Lake Tanganyika's fish populations, already under heavy fishing pressure outside of the Park's waters. As such, Mahale Mountains National Park provides an important contribution to sustaining the Lake's fishing industry—a vital source of protein and income for the majority of community members living along the lake shore as well as further afield.

Since fishing is the backbone of socio-economic activities in the communities nearby, the park preserves the breeding places for fish as a method of preserving the fishing business to support the lives of people around the lake. Most of these fish are endemic and can only be found in Lake Tanganyika. For wildlife-based tourism, the fishing industry is as significant as endemic and exceptional resources; wildlife visitors may use the fish for consumptive and non-consumptive purposes.

Moreover, providing community services to residents surrounding the park is another socio-economic advantage. The Mahale Mountains National Park delivers many advantages from its continuous conservation efforts. The Community Conservation Service frequently extends the park's financial and non-financial advantages to the communities surrounding it. There are a lot of career options for residents near Mahale Mountains National Park.

Cultural Resource Values

Mahale Mountains National Park's general management plan views cultural resource values as the second category of exceptional resources. In this category, all values connected to cultural resource values inside the protected area are safeguarded and lawfully exploited.

In developing nations, certain protected areas were created in the ancestral locations of neighboring populations. The Mahale Mountains National Area is no exception, as villagers are permitted to come and practice their spiritual rites there. The overall management strategy emphasizes that.

> Tongwe were the first people to populate the Mahale peninsula and their culture is tightly linked to the natural landforms within and around MMNP. The Tongwe believe many of their ancestral spirits are found in the various mountains, rivers, and wetlands of the Mahale region. In the past, the Tongwe used many mountain sites for prayer and sacrifice to ancestral spirits t are now located within MMNP.

According to the management plan, the Tongwe people have significantly safeguarded many portions of the forest via the taboos and management constraints linked to these holy locations. The management plan addresses some of the fundamental needs of sustainable development by directly incorporating locals' traditional knowledge into conservation.

Giving the local people access to outstanding resources and employing their traditional knowledge for conservation creates a bridge to ideas of sustainability. In this stage, following the successful completion of library or field data collection, the planning team, along with other stakeholders, may apply SWOT to analyze the data into meaningful and user-friendly interpretations.

At this time, the team should correctly define what the protected region is, what the exceptional resource values are, and finally, its importance. To establish the protected area's uniqueness, the team may compare the acquired information to that of local, regional, and worldwide protected areas. Let's look at this example:

> Lake Tanganyika is the longest in the world (720 km) and Tourism at Mahale Mountains National Park.

The highlighted remark was obtained from the general management plan of Mahale Mountains National Park; it only shows what has been mentioned in the preceding paragraph. After investigation and inventory, Lake Tanganyika is technically considered to have both scenic and socio-economic significance.

A comparison study on strengths, weaknesses, possibilities, and dangers revealed that Lake Tanganyika is the longest lake in the world because it was purposefully compared to other lakes at local, regional, continental, and global sizes. In short, the presentation on the interpretation of the acquired data extends our discussion into the formulation of vision, aim, and management goals.

With exceptional resource values clearly defined, classed, and recorded, the next stage is determining the management goals, as strengths, weaknesses, threats, and opportunities in each exceptional resource value have been recognized and documented. In other words, the management goals aim at developing the strengths, improving the weaknesses, reversing all risks, and unlocking the possibilities in each recognized extraordinary resource value.

When defining the vision and goals, the protected anticipate the condition, status, or appearance of protected areas over the following 10 years. Based on the existing status of exceptional resource values, the protected areas define long-term objectives or goals to be fulfilled after the general management plan.

In some plans, this future is specified as short-term goals or ambitions of the protected region; in others, it is specified as long-term aspirations. In some more recent plans, it has taken the form of a vision statement. In such circumstances, the general management plan should include a vision statement, commonly referred to as a purpose statement, as required in the Mahale Mountains National Park:

Park Purpose Statement of Mahale Mountain National

> To protect a sizable Albertine Rift ecosystem and its globally significant biodiversity, especially the chimpanzees and other primates, the endemic fish, and the mosaic of vegetation and habitat types.

With the purpose statement and management goals in place, the next stage is to work with the stakeholders to build the strategies for attaining the vision via management objectives. As there are typically various methods by which this may be done, the range of alternatives for management activities should be defined, and the right ones should be picked.

Management activities may include management zoning or buffering when the protected areas are further partitioned into distinct management zones. In this context, management zones are specifically defined and delineated regions or zones with varied management and usage regimes, including activities, authority, resources, and constraints within the protected area.

This is one of the methods of sustainable usage and protection of wildlife resources, where access to each zone may depend on time, resilience, carrying capacity, and fragility of the animal resources, among other factors. As a result, some of the regions are referred to as wilderness zones, low-usage zones, high-use zones, hunting zones, and no-use zones, among other things. The size and number of management zones are defined by the stakeholders.

In brief, the management zones provide a framework for meeting and resolving the dual management requirements of protecting the park's natural features and environment, as well as regulating and encouraging tourist usage. Management zones designate places where comparable kinds and degrees of use and management emphases are applied based on the vision of the protected areas, their exceptional resource values, acceptable park uses, and management goals.

The 1996 Serengeti Management Zone Plan defined eight zones; however, they eventually proved impracticable at the park management level. As a result, the following general management plan streamlined the zoning structure into high-use zones, low-use zones, and wilderness zones.

With clearly designated and physically delineated management zones, the management goals for each exceptional resource value are further converted into concrete tasks. It is crucial to note that our discussion on the relevance of the general management plan emphasized that the plan assists in the monitoring and assessment of the efficiency and effectiveness of the protected area's management.

That is to say, during planning, each action may have a specified time, location, responsible person, and resources necessary for execution. With all this sort of information, the monitoring and assessment of the strategy become pleasant, open, and uncomplicated.

The documentation of all exceptional resource values, vision or purpose statements, goals, and activities drives the planning team to integrate and compile all of the aforementioned planning parts into a single document known as a draft of the general management plan.

The draft gives the possibility for another level of public stakeholders' participation. Stakeholders are usually interested in discovering whether their thoughts and recommendations were effectively reflected in the suggested general management strategy. Recognizing that a draft is not a final general management plan is critical because it is always subject to subsequent changes and revisions.

In that scenario, public consultation is the phase that allows stakeholders, or the general public, to examine the draft management plan. It is an essential step in the planning process, as it promotes trust among the stakeholders. Public consultation takes place in a variety of formats to suit the opinions of consulted stakeholders, including general meetings, targeted group discussions, and social media. It also takes on people with unique requirements, such as being handicapped.

Following public consultation, the draft is amended, taking into account the opinions of stakeholders and the public. Good practice mandates that all written and spoken comments received, including those acknowledged during public hearings, should be documented and examined. The inclusion of all comments and correct editing of the documents make the plan suitable for the next phase, which is approval of the general management plan.

The plan may move through different organizational decision boards before it becomes a comprehensive general management plan that is ready for execution. When the management plan has been produced and authorized and the operational plans are in place to guide its execution, field staff are then able to put the plan into effect.

It is a crucial phase as it ensures desired outcomes are realized, enhances and supports the administration of protected areas, develops common understanding, produces new information, motivates stakeholders, assures accountability, and encourages public and political support.

The last phase in the planning process is to decide on either a review or update of the management plan. In many circumstances, the plan will be time-limited by law, often for 10 years in Tanzania. An important aspect of this last phase is to ensure that monitoring input is provided to aid in the formulation of the new or future general management plan. The protected area may hire a dedicated person to monitor and analyze every action in the management plan. It is vital to see this period as a learning phase.

Human Dimensions for Sustainability

Wildlife tourism attracts both domestic and international tourists due to the unique behavioral and structural features of wildlife resources. Despite the resources' uniqueness, it is critical to understand wildlife tourism's motivations, expectations, and satisfaction when it comes to marketing wildlife destinations.

A thorough understanding of tourists' motivations, expectations, and satisfactions may lead to the customization and enhancement of wildlife tourism products and services. The business-as-usual scenario in the execution of wildlife tourism entails that wildlife tourists are highly satiated with the behavioral and structural features of wild animals, though, in a real sense, wildlife tourists need and engage in more than what most tourism experts comprehend.

Wildlife visitors may appreciate and support wildlife in numerous ways to enhance its productivity and existence for the benefit of current and future generations. The way wildlife travelers are portrayed sounds like they rarely have any positive values for wildlife species and protected areas.

More regrettably, their appreciations for wildlife are hardly acknowledged and publicly declared, particularly among the normally traveled travelers whose political, educational, and social identities are not well known. In most instances, tourism and conservation managers regard wildlife tourists as agents of destruction and therefore need close surveillance and stringent regulation to reduce their anticipated adverse impacts.

Because of the unpredictable nature of wildlife tourists' behaviors, conservationists and animal activists are sometimes doubtful about their support for and appreciation of wildlife species and habitats. For example, wildlife species have been recently treated as photo props, some wildlife tourists hardly advocate for wildlife, some wildlife individuals have already been converted into humans, and informal feeding of wild animals is a normal occurrence.

Also, support for wildlife conservation remains a responsibility of conservation agencies and local people; wildlife tourists and operators hardly take time to research what they eat, what they buy, and where they visit, and wildlife tourists and operators hardly support the local community's socio-economic development activities.

With the advanced development of camera and mobile phone technologies, some wildlife visitors have been squandering the opportunities they get from photographic and selfie tourism. As travelers move to areas where they can easily snap photos with animals.

In some circumstances, this may entail the chemical capture or sedating of wild animals for wildlife enthusiasts to take portrait photographs. It is a novel niche of wildlife tourism that satisfies the interests of special visitors. Using untamed animals as photo objects for decorations is not well documented, publicly declared, or marketed, but it does exist.

It is not terrible if we also take some time to grasp the adverse consequences or the other side of using animals as decorations or social media entertainment. Capturing wild animals for any purpose entails several unpredictable dangers and is hazardous to both wildlife and travelers.

Depending on the type of wild animal, docile or non-docile, if imprecisely executed, chemical capture may kill or severely injure the animal and the wildlife tourist. Additionally, it is legal and ethical to take animal photos in their natural environment without interfering with their ecological or social activities.

To support or appreciate the role of wildlife, wildlife travelers should also speak out against wildlife injustice and mismanagement. Wildlife travelers are of every professional description, including lawyers, conservationists, physicians, journalists, researchers, and justices, among others.

With their professionals, commitments, interests, and abilities, they can speak for wildlife against any type of injustice. Wildlife has been subjected to violent predation, habitat degradation, trafficking, and tormenting. If everyone remains silent, all of these unwanted acts have the potential to drive all wildlife species to extinction.

Some wildlife tourists are actively involved in poaching and trafficking offences, or traffickers fall under the umbrella of wildlife tourists. Sadly, some tour operators, wildlife tourists, and guides may sometimes be aware of all these crimes, but for undetermined reasons, they don't either notify conservation authorities or halt the poaching intentions.

Wildlife travelers should advocate for wildlife to save the wildlife and their habitats when they notice anyone illegally capturing, tormenting, and murdering wild animals or accumulating anything from protected areas.

Wildlife travelers and operators never vacillate or shrink away from speaking up for wildlife in order to save unique and finite resources for current and future generations. It is one method of appreciating the local, national, and international conservation efforts of the wildlife resources a visitor has just visited.

Without such efforts, wildlife travelers would have never seen such gorgeous animals and vegetation in a natural setting. The essential point to remember is that the uniqueness and beauty of all wild animals, wild flora, and their naturally occurring environment in protected areas come with enormous expenses.

It would be very simple to comprehend the previous statement if I at least cited the following concerns: did you know that some people whom we call rangers get murdered every day by armed poachers when safeguarding wildlife? Are you aware that dangerous wild animals, including venomous snakes, hippopotamuses, and African bison, murder and severely injure our conservationists every day while protecting them against poachers?

How many conservationists lose loved ones and marriages to protect wildlife? Now, why can't wildlife tourists and operators just speak for wildlife by disclosing any threat against it? Why don't they at least support the interminable, tiresome, and hazardous conservation efforts? Paying the entrance fee is one, but not the only, method.

Surprisingly, some individuals from wildlife species have been or are about to be given human names. In some countries, including Tanzania, it was officially proclaimed that anyone who wishes his or her name to be allocated to an individual untamed animal should pay a certain amount of money.

In that context, I am personally uncertain if the natural or untamed beauty and originality of the named individuals are or will still be psychologically valid. Perhaps this requires an extensive description. According to the responsible conservation authority, a destitute person's financial muscles determine the allocation of a name to an individual wild animal.

In other words, you pay the money and then get the animal names. Although nobody knows whether the authority scrutinizes the type of name before giving it to the untamed animal, it also has a mechanism in place to check where and how the payees get money. More importantly, the number of names and animals that a payee can supply.

There are four main concerns here. First, what if the person chooses to designate the unusual moniker for an individual who disrespects the untamed animal? What if the recipient earned money from poaching or any illicit wildlife business? What if the payee assigns a name that leaves tourists with a terrible memory? What if naming only implies more responsibility, conservation priority, and protection for the named individuals?

Based on the expressed concerns, it is evident that supporting and appreciating wildlife should not only be for named individuals but also for everyone, including unnamed, unknown, and embryonic ones. It is unjust for wildlife tourists, operators, and conservationists to support and appreciate certain individuals or species because of their long-term and regular documentation, attractiveness, and abundance.

Similarly, it is imprudent to disregard the species that are challenging to conserve due to heightened conservation costs and frequent problems with neighboring communities. Additionally, the introduction of wildlife tourism presents a new challenge to wildlife conservation.

However, many protected areas still rigorously discourage the act of wildlife feeding. Wildlife travelers' involvement in feeding animals, both legally and illegally, introduces several behavioral and ecological problems for wildlife. Wildlife travelers do not need stringent regulations and a regular awareness program to comprehend the impacts of casual wildlife feeding, but their support is needed to stop this. Wildlife poisoning, disease transmission, and the loss of wildlife's natural behavior may cease with their support.

Wildlife tourists and operators rarely research their buying, eating, and visiting habits. As a result, wildlife visitors engage in unlawful wildlife business, such as purchasing souvenirs from rare, endemic, and endangered species. Also, because they don't enquire about or investigate whatever they do at the destination, even the food they consume may be unlawfully collected from protected landscapes.

Moreover, many wildlife travelers are completely unaware of the fragility of the areas they visit and the extent of the scarcity of resources they use. For example, access to water, sustenance, and reproductive areas may be a major problem for most of the fauna. Furthermore, wildlife visitors, appreciating wildlife conservation, and the socio-economic development of local communities are the responsibilities of wildlife visitors and operators.

Responsible wildlife tourism should be an alternative source of conservation and community development funds. Wildlife tourism does little to support and appreciate conservation and socioeconomic development initiatives. Wildlife visitors hardly consume locally prepared or grown foods, stay in locally owned accommodation establishments, and do not contribute any funds to conservation activities. In those regards, wildlife tourism discourages conservation agencies and local communities from engaging in wildlife tourism.

Expectations of Wildlife Tourists

The expectations of wildlife tourists entail all those aspirations and desires every tourist would like to fulfill after visiting certain wildlife tourist destinations. The essential point here is that travelers have been working for years, casually and formally, days and evenings, to save money, hoping that one day they will visit a destination of their desires.

Some tourists would like to visit and climb Mount Kilimanjaro or Mount Everest, while others would like to see the wildebeest migration in Serengeti National Park, the tree-climbing lions in Lake Manyara National Park, or the natural flowers in Kitulo Plateau National Park. All of those are the aspirations, desires, and expectations of wildlife travelers.

More importantly, wildlife visitors would like to visit places where they can physically interact with untamed animals and vegetation so that they can increase their knowledge of wildlife and nature in general. In this era, environmentally conscious travelers would like to learn more practically about untamed animals and nature.

They would like to comprehend nature conservation and, if possible, contribute to the continued efforts of wildlife conservation undertaken by local communities, conservation agencies, and international communities at large. Also, wildlife tourism would like to have environmentally, socially, and economically responsible, secure excursions that respect the well-being of local communities, wildlife, and themselves.

Many visitors are against the excursions that violate conservation laws and the rights of indigenous people. Tourists would expect excursions that respect the untamed animals and the ecosystem as a whole. Engaging in harmful, unethical, and negligent excursions that violate wildlife laws, such as off-road driving, feeding wild animals, and driving many vehicles around one animal, may disappoint many wildlife tourists.

All wildlife tourism stakeholders must participate in safe, equitable, transparent, and ethical wildlife excursions that respect the privacy and rights of tourists, indigenous people, and wildlife. Otherwise, the stakeholders may disappoint and deceive wildlife tourists, which is unethical and goes against the principles of sustainability.

Most wildlife tour operators can hardly engage in responsible wildlife tourism operations because they are expensive and demanding in terms of time, ethics, and other resources. Wildlife tourists expect to engage in wildlife tourism where wildlife is really free-ranging, healthy, and well-preserved, and local communities that incur conservation costs are well taken care of.

Many wildlife visitors will be immensely discouraged after visiting protected areas that lack sufficient connection to other protected areas. Animals can't move or reproduce freely, and creatures do not have access to some of the ecological resources. Also, the areas where local communities have been denied access to essential fundamental requirements because of the fauna. More regrettably, human-wildlife conflicts are monotonous, and there are no methods to resolve them.

Furthermore, some wildlife enthusiasts would like to observe the wildlife in their natural environment, expressing their natural behaviors. In the introduction section, I have described explicitly what it takes for an organism to be a fauna. Those are the features that visitors would like to witness when visiting protected areas.

Without such qualifications, travelers would be discouraged and disappointed, as what they expected was not there. All of the protected areas and natural resources must be maintained as naturally as possible.

In summary, wildlife tourists expect quality wildlife resources as routinely advertised and presented in the tour safari package. Wildlife tourism destinations, like other tourist destinations, tend to market themselves using various strategies to expose their unique resources to prospective tourists.

In the endeavor of marketing, the destination constructs its image and therefore many expectations from wildlife visitors. Unfortunately, every prospective tourist may have a unique image and expectations of the destination. The expectations may differ based on whether they are solely wildlife- or non-wildlife-related.

In the case of wildlife-related expectations, wildlife travelers may end up asking themselves several questions about the wildlife resources at their destinations. Depending on the sources of the information, in this context, I mean the marketing channel the prospective tourist obtained the destination information from, as some tourists may enquire about the authenticity, actuality, and actual existence of marketed wildlife resources at the destination.

We are in the digital era when information can be falsified or manipulated to attract wildlife travelers' perspectives, attention, expectations, and desirability. Like any other business, wildlife tourism can be geographically or resourcefully falsified to attract many tourists to the incorrect destinations.

In some destinations, wildlife resources are misrepresented as tourism products and services. It can be demonstrated in public or private protected areas to attract a wide range of wildlife visitors. The more travelers you have, the more revenue the protected areas generate. In this business model, the protected area may provide incorrect information on the presence and uniqueness of resources that do not exist to attract many visitors.

Unfortunately, destination managers have always had strategies for getting away from falsehoods about the absence or unattractiveness of resources. Moreover, some wildlife destinations may falsify the geographical location of the resources through the declaration of wildlife resources from other protected areas or the protected areas of other countries as part of their wildlife resources.

A neighborhood of protected areas or countries usually attracts the protected areas or countries with insufficient and uninspiring resources to obtain the geographical advantage of the other protected areas or countries with magnificent wildlife resources. In those circumstances, some prospective travelers always need to clarify all uncertainty about their prospective wildlife destinations before making decisions.

Wildlife managers must avoid such unscrupulous suspicions by using proper tourism marketing platforms, such as embassies, magazines, people, and digital media, to acquire the trust of prospective tourists. Also, to provide accurate resources and geographical information, some travelers may have previously visited destinations where the difference between the advertised and actual wildlife attractions was significantly prominent.

However, the doubtfulness of perspective among wildlife tourists occurs due to the rarity and uniqueness of the marketed wildlife resources, which makes it difficult for wildlife tourists to believe that such resources and beauty exist in this world.

As an illustration, consider the presence of more than two million migrating wild animals between Serengeti National Park and Masai Mara National Reserve.

Imagine witnessing more than two million undomesticated animals in one place while simultaneously migrating. It may sound like a digital exaggeration. Though it is true, it sometimes does not make sense, particularly for those who have never been to the area. In this era of abundant and readily accessed digital and analogous information, many wildlife visitors may search for reliable information before and after their visitations.

Websites, television, periodicals, and acquaintances are all sources of information. Prospective visitors make these efforts to ensure that the wildlife resources meet their expectations. The correct descriptions of the resources should match the tangibly available wildlife resources as an alignment of wildlife tourism to the objectives of sustainability. Any falsification or deception in wildlife tourism may tarnish the destination's or country's image, as well as demoralize wildlife tourists, as their expectations are usually never fulfilled.

Moreover, prospective wildlife visitors have many expectations for non-fauna resources. Wildlife tourist destinations are expected to have suitable infrastructure, transportation, superstructure, amenities, and safety for a seamless wildlife tourism experience. Apart from distinct and abundant wildlife resources, the destination should offer affordable and convenient non-wildlife resources to facilitate the involvement of various sectors and enhance the experience of wildlife visitors.

As humans, wildlife visitors anticipate consuming, drinking, and slumbering while in wildlife tourist destinations. In this case, the availability of adequate, affordable, and convenient food, beverages, and accommodations is a matter of concern. Wildlife travelers expect a variety of food, beverage, and accommodation services with varying qualities and prices.

Variations in quality and price may accommodate the needs and capabilities of diverse wildlife travelers. In the case of accommodation, some wildlife visitors may prefer traditional to modern accommodation establishments, while others are constructed within or near protected urban areas.

Wildlife travelers also expect their destinations to have a variety of transportation infrastructure and facilities to support their movements within and between them. Without reliable and integrated conveyance systems, wildlife travelers could hardly access wildlife destinations.

The availability of air transportation infrastructure is necessary for wildlife tourism development. International wildlife visitors expect affordable and reliable air transport to conveniently and easily access tourist destinations; in a short time, air transport may cover the extended travel distance between tourists' residences and destinations.

More importantly, the closer the air transport is to protected areas, the more convenient it is. The majority of international airports in many countries are located far away from the respective protected areas; some protected areas have airstrips connecting them to the main airports. As a matter of convenience, such airports serve as centers for chartered and scheduled air transport to protected areas.

However, in developing countries, air transport is still regarded as a luxury form of transport, afforded mainly by travelers with enormous financial resources. Furthermore, many tourists believe that local air transport in developing countries is hazardous and unprofessionally managed by incompetent airlines and personnel. Lack of ability to prevent avoidable accidents and frequent airport delays are some of the factors that justify their accusations.

Despite the ongoing inconveniences caused by hostile infrastructure and travel facilities in some destinations, wildlife visitors would prefer surface-to-air and water transport because of their convenience and affordability. As a result, many wildlife travelers expect their destinations to have appropriate, affordable transport networks for accessing all wildlife resources.

Many protected areas, particularly in developing countries, are geographically located in remote regions where access is only or mostly possible via road or railway transport. Wildlife visitors travel for hours before gaining access to those resources. At the end of the book, some travelers describe it as an adventure.

Furthermore, wildlife travelers expect a variety of amenities, including financial transactions, entertainment, telecommunications services, medications, and many more. Wildlife tourist destinations should be able to cover all of these amenities to enhance the experience of wildlife tourists.

As vacationers expect to routinely communicate with their relatives and acquaintances when visiting destinations, reliable, affordable, and convenient telecommunications services are essential. To enhance visitors' travel experience, internet connection, and telecommunication services should be implemented around accommodation establishments and protected areas.

Access to their bank accounts is equally important for all tourists to conduct financial transactions at their destinations. Mobile and Internet banking, for example, can improve the payment of the entrance fee to protected areas, as well as the purchase of souvenirs.

Furthermore, the development of transport infrastructure is equally essential, as in some protected areas, certain wildlife tourism activities are impossible because of inadequate infrastructure. Tourists expect every wildlife tourism destination to have a supportive infrastructure so that they can appreciate the distinct wildlife resources.

Unfortunately, the majority of international wildlife visitors come from the developed world, which has extensive infrastructural development. When visiting developed countries, visitors expect the tourist destinations to have similar infrastructure to their own countries. Unfortunately, infrastructural development within the protected areas is sometimes limited by either the budget or policies of the protected areas.

To clarify, some protected areas barely consider the construction of tarmac roads, while others allow it. Last, but not least, wildlife visitors expect the protected areas and the governments to support local businesses and communities that have been evolving or coexisting with wildlife.

Since wildlife tourism is the product of wildlife, and local communities are at the core of conservation, wildlife tourists expect wildlife tourism to benefit local communities. When discussing the sustainability of wildlife tourism, I also highlighted that nowadays, the majority of wildlife tourists are environmentally and socially cautious. Most travelers would also like to know where their money goes, and they understand that communities are responsible for wildlife resource conservation and management.

Satisfactions of Wildlife Tourists

During an excursion, wildlife tourists usually make general or specific judgments and evaluations regarding wildlife attractions, tourism services, and products. It helps to understand wildlife tourists' overall satisfaction with the excursion. In the world we live in, the pursuit of a stress-free life and contentment are vital. Everyone needs a joyful and stress-free existence.

Wildlife visitors equally need a joyful life that can be measured not only by the amount of money they spend and the locations they visit but also in different ways. I remember visiting "Stress-Free Beach," one of the beaches on the Tanzania side of Lake Victoria.

It may be the aim of the beach to get rid of all burdens from their customers, or, in other words, to make their customers happy to the extent of forgetting most of their problems, but how the beach makes them happy could be another topic.

Similarly, how does wildlife tourism make visitors joyful or tension-free? Wildlife tourism gratifies wildlife tourists in a variety of ways, including by making them retreat from daily routines, social bonding, learning about nature, exhilaration, and engaging in adventure activities.

When precisely and timely practiced, wildlife tourists can get rid of many tensions and finally become joyful. When wildlife visitors are pleased, it means her expectations have been undoubtedly met.

Wildlife tourism presents situations where wildlife tourists can escape from the daily activities that are always distressing and dull. It exposes travelers to a new and unfamiliar environment and allows them to experience new modes of existence distinct from their residences. Wildlife tourism expeditions unwind by savoring the beauty and scenery of various natural and physical features endowed with protected areas.

In Chap. 5, cascades have been described as one of the methods wildlife tourists can use to get free of stressors, which worrisome wildlife tourists can use. The pure, natural, and aromatic air from the natural vegetation, as well as the irrepressible melodies of birds and insects, make wildlife tourism in the protected area peaceful and joyful.

Additionally, wildlife tourism has been well-known for decades for reviving and strengthening social bonds among wildlife tourists. As friends and relatives travel together for wildlife tourism expeditions, this can also be an occasion for either visiting friends and relatives or strengthening family connections.

Wildlife travelers travel miles and miles from their residences to also visit their friends at their destinations. This is prevalent in developing countries, where the diaspora from the developed world may visit wildlife attractions with their families in mind.

Similarly, privacy and relaxation during wildlife tourism expeditions may help wildlife tourists strengthen or revive their family bonds. As a result, many wildlife tour operators offer special family itineraries, such as the Honey Moon Package, for wildlife tourists to enjoy with their families.

Moreover, wildlife tourists become satisfied after learning about the underlying behaviors and interactions of wildlife species in the natural environment. Apart from appreciating the grandeur and diversity of wildlife resources, responsible visitors would also like to learn about ecology, biological diversity, local culture, and their relationships with wildlife conservation.

Responsible visitors are generally curious, as they would like to grasp many aspects of wildlife conservation in each protected area they visit. In that case, the wildlife tour operators must have competent tour guides to satisfy the requirements of responsible visitors by professionally explaining all concerns raised. Such travelers are curious and more knowledgeable about environmental and cultural issues, as they always compare the provided information to other locations.

Furthermore, wildlife visitors have varied originality; some would be content if engaged in physical wildlife tourism activities, while others would need gentle ones. It is essential to expose adventure tourists to physical and hazardous tourism activities such as hunting, camping, mountaineering, rafting, and diving.

Engaging in those operations may make them joyful and fulfill their desires, as most of them are young and lively. Non-adventure tourists, on the other hand, are generally pleased when engaged in gentle wildlife tourism activities such as game viewing and forest dinners, which usually require little energy.

Motivations of Wildlife Tourists

In the context of motivations, many wildlife tourists are exclusively motivated by naturally occurring wildlife resources rather than confined or restrained animals, as naturally occurring species demonstrate natural features that many wildlife tourists have never witnessed before.

The display of movement, grazing patterns, and reproductive behaviors in the natural landscape motivates their engagement and expenditure on wildlife tourism activities. It is alluring to see untamed natural flora and animals whose behaviors are not directly or indirectly influenced by humans.

To summarize, motivation among wildlife travelers may vary depending on the visitor's interests. As various visitors have various interests in different fauna species, for better comprehension, we can analyze the interests of various wildlife visitors based on the terminology of wildlife, just as we explicitly discussed in the introduction section.

Remember, we explained that wildlife is composed of various components: living and non-living, vertebrates and non-vertebrates, untamed plants and animals, and their naturally occurring environment. From such components, it is probable to pinpoint the interests of various wildlife travelers.

When discussing planning in the protected areas, we talked about scenic, physical, and cultural exceptional resource values. In other words, these are non-living objects located in protected areas that attract various visitors. Such features include water, slopes, mountains, reservoirs, devotional areas, and monuments.

Some visitors are extremely motivated by the aquatic bodies inside the park for canoeing, recreational fishing, and water exploration. Some travelers are also interested in climbing mountains in protected areas.

Wildlife visitors would prefer to visit protected areas with distinct historical and cultural sites; wildlife visitors would like to visit such areas for educational and rewarding reasons. Some visitors visit protected areas, such as Ngorongoro Conservation Areas, where natural, cultural, and historical resources meet, just to admire the historical and cultural resources.

Moreover, the protected areas with worship sites, as well as the ancestral areas of the local communities residing around the park, can be an important destination for culturally sensitive visitors. In developing countries, many protected areas have worship areas for local communities inside the park where local people routinely go for worship.

Wildlife tourists are only interested in living resources in protected areas, which in most cases are untamed flora or animals. Wildlife visitors would merely appreciate the diversity and majesty of untamed flora and animals. The morphological and behavioral characteristics of untamed vegetation and animals are educational and entertaining. Moreover, wildlife travelers would like to learn more about natural flora and animals, their interactions, and ecological values.

However, despite the focus on wild plants and animals, some wildlife visitors are more interested in wild plants than animals. Such travelers would like to visit protected areas with a high diversity of vegetation species, such as forests and nature reserves.

In protected areas, wildlife travelers relish and appreciate the exceptional structural and behavioral features of untamed plants. The length and size of trees, the blossoming of wild plants, the aromatic scent, and ecological features such as water towers and canopy sizes are some of the fascinating and educational features motivating wildlife travelers to visit the protected areas.

Some wildlife visitors are exclusively interested in untamed animals. Tourists would like to experience every aspect of the animals, from feeding, locomotion, mating, foraging, and their social activities. Visitors would appreciate it and learn a lot from observing untamed animals displaying their behaviors in natural landscapes.

However, the speciality in untamed animals may be subdivided further into other components. As a result, some wildlife tourists are more motivated by birds than other wild animals, particularly melodious and attractive birds, because they are fascinating to interact with.

The trophy hunters are only motivated by the presence of five large mammals because an enormous trophy is their only motivation. In that case, trophy hunters will visit public and private protected areas with robust and prominent species for hunting. Other wildlife visitors are motivated by the dolphins' desire to swim with them.

It makes such vacationers appreciate and feel special when swimming with dolphins in their natural habitats without any damage. This is referred to as a hydro, or water adventure, where the vacationers swim at their own risk. Swimming with dolphins and trophy hunting are adventures because travelers sometimes like to pursue challenges and exhilaration.

Supporting local communities and conservation initiatives motivates wildlife visitors to visit the protected areas. Socially and environmentally responsible travelers foster sympathy for wildlife conservation efforts as well as, more importantly, the roles local communities play in conserving wildlife species.

In their efforts to provide help, responsible wildlife visitors are highly motivated by supporting the local communities and wildlife conservation agencies. Contributions of financial and non-financial resources are the common contributions that wildlife visitors always make.

Chapter 7
Responsible Wildlife Tourism

Abstract This chapter provides an in-depth overview of responsible wildlife tourism. As a continuation of sustainable wildlife tourism, responsible wildlife tourism principally emphasizes the practical application of ethical, sustainable, and conservation principles for every wildlife tourism stakeholder. Since wildlife tourism continues to expand while providing unique and outstanding tourism products and experiences to wildlife tourists globally, it is imperative that every wildlife tourist becomes responsible for safeguarding the fragile wildlife attractions and destinations. Unregulated mass tourism has frequently led to substantial detrimental social, cultural, and environmental damage worldwide. Responsible tourism aims to address some of the challenges of responsible tourism by encouraging ethically sound, socioeconomically, ecologically, and culturally acceptable approaches that prioritize wildlife and world well-being first. Responsible wildlife tourism originates from sustainability, which focuses on safeguarding wildlife resources for present and future generations by addressing intergenerational and intragenerational equity. It reflects the concepts of sustainability by educating tourism stakeholders to be responsible for their behaviors. Conservation agencies, travel operators, local communities, and wildlife tourists are working together to formulate and implement sustainability policies for the enhancement of responsible behaviors such as avoidance of off-road driving and feeding wild animals, empowering local communities, and respecting Indigenous culture. Responsible wildlife tourism emphasizes ecological restorations and protection of endangered species while enhancing financial and capacity-building support for the conservation of protected areas. It supports and promotes community involvement through traditional activities, employment, and conservation, which are crucial in advocating and executing social justice and cultural preservation for Indigenous populations. Responsible wildlife tourism is vital in ensuring the highest level of tourist satisfaction while reducing ecological disruptions and enhancing interactions between wildlife tourists and the local community. Responsible wildlife tourism advocates the adoption and employment of environmentally friendly technologies such as solar-powered motor vehicles and accommodation facilities. The implementation of such facilities and sophisticated monitoring equipment can alleviate the detrimental impacts of climate change and greatly reduce ecological footprints.

A. A. Mamboleo, *Sustainable Wildlife Tourism*,
https://doi.org/10.1007/978-3-031-97795-4_7

Keywords Tourism carrying capacity · Environmental education · Community involvement · Ethical tourism practices · Technological innovations in tourism · Post-pandemic tourism resilience

In the previous chapters, the author highlighted integrating wildlife tourism with the principles of sustainability. The entire concept of sustainability rests on the importance of safeguarding and conserving wildlife resources for the benefit of current and future generations, which is the responsibility of wildlife tourism stakeholders.

As a guiding policy, the principles of sustainability strive to remind tourism stakeholders that the offering or consumption of every tourism service or product should respect the environmental, cultural, and economic dimensions of wildlife resources. Similarly, the shifting of wildlife visitors and wildlife operators from conventional to responsible ones is equally essential.

Some of the features of responsible tourism as the behavioral implementation strategy of sustainable tourism include the substantial minimization of the adverse impacts of conventional or mass tourism. Mass or conventional tourism has calamitous environmental, economic, and social impacts due to massification and inadequate guidance, as previously described.

Since mass tourism is a profit-orientated industry, tourists and tour operators hardly consider the other side of their activities. As part of the strategies for turning conventional into responsible wildlife tourism, tourism stakeholders must purposefully regulate the objectionable behaviors of tourists by reducing the number of tourists visiting a particular area at a particular time.

It is also tourism-carrying capacity, a concept derived from wildlife ecology to preserve the environment. With this concept, explicit guidelines and regulations may limit the number of tourists visiting an area, as well as, ultimately, the associated impacts of wildlife tourism on protected areas, communities, and the economy.

In that case, each wildlife attraction or destination may have specific regulations and guidelines for implementing sustainable wildlife operations. Furthermore, the guidelines and regulations should specify the number of visitors, as well as all, permitted and prohibited activities that may prevent travelers from threatening wildlife resources.

Conservation agencies and destination management organizations are also responsible for formulating and implementing appropriate strategies for monitoring and regulating wildlife visitors' behaviors. The regulation should further specify the possible penalties for violation of any of the regulations, which may discourage improper refuse management, unlawful animal feeding, off-road driving, and pursuit of the animals.

The other side of responsible wildlife tourism is maximizing the adverse impacts of conventional tourism. Supporting conservation initiatives is one of the essential aspects of wildlife, as some wildlife visitors are highly motivated to contribute financial and non-financial resources to conservation authorities to support species and landscape conservation initiatives.

Responsible visitors and wildlife tour operators' efforts are directed toward supporting fragile habitats and endangered and uncommon species. Moreover, the contributions may occur in the form of technical empowerment, human resources, financial resources, technologies, and apparatus.

More importantly, responsible wildlife tourist destinations, protected areas, wildlife visitors, and wildlife tour operators support conservation and development initiatives in local communities adjacent to significant wildlife attractions. It is the same communities that routinely endure the costs of wildlife conservation by contributing their lands, time, properties, and lives to it.

In an endeavor to moderate and encourage the local communities to continue supporting wildlife conservation, responsible wildlife stakeholders provide support at institutional and household levels with social infrastructure such as the construction of schools, roads, dispensaries, and local markets.

Sustainable wildlife tourism, as a means of enhancing social equity, recognizes the rights of marginalized groups by involving them in the conservation and use of wildlife resources. Sustainable wildlife tourism should respect the rights and interests of marginalized groups and, particularly, local communities.

There have been instances where local communities are allowed to enter protected areas for veneration, the accumulation of non-timber forest products such as traditional remedies, and the production of honey. Moreover, locals are trained, empowered, and employed at the wildlife attractions.

Additionally, sustainable wildlife tourism usually attempts to provide quality tourism products and services to gratify wildlife tourists. It is a method of doing justice to wildlife tourism by valuing their money, time, and reverence for wildlife resources. The satisfactions of wildlife tourism have myriad benefits for the destination and protected area itself, including collateral effects where tourists may also visit another protected area due to established trust.

Just as described before, provisional quality wildlife resources and memorable experiences may elevate the destination's reputation. In that sense, sustainable tourism is equally responsible for providing quality services to wildlife visitors while taking care of their safety and security. It should ensure that the provided services are equal in value to the money and time spent by wildlife visitors.

Furthermore, sustainable wildlife tourism takes place in regions where endangered species and ecologically sensitive areas are the primary priorities. Since environmental education is one of the major characteristics of responsible wildlife tourists, multi-tasked protected areas—recognized as breeding grounds, refugia, water towers, migratory corridors, and dispersal areas, among others—may provide unique opportunities for responsible wildlife tourists to learn and understand without contradicting environmental integrity.

Understanding the importance and status of such protected areas may make travelers more responsible and disseminate their acquired knowledge to other tourists. Additionally, this may further convince them to be more environmentally responsible and contribute their resources to conservation initiatives.

Furthermore, sustainable wildlife tourism supports and entertains well-planned, monitored, guided, and self-guided tour packages. On those excursions, respecting the privacy and integrity of untamed animals is inevitable. Privacy necessitates respecting untamed animals when grazing, stalking, or mating.

Some travelers, whether on guided or self-guided excursions, drive off the designated routes in search of the most attractive wild animals; lions can be one of them. Furthermore, some travelers may interfere with hunting or untamed animals to satisfy their desires. Some travelers may even bribe the operators to make sure they see lions on their honeymoon.

Such tourists' actions may disrupt the natural behaviors of untamed animals. Responsible travelers and tour operators may report objectionable behavior to authorized authorities, which helps to prevent and reduce such harmful acts. Also, the adoption and utilization of appropriate and environmentally favorable equipment on wildlife tourism expeditions makes wildlife travelers and tour operators environmentally responsible.

In protected areas, some wildlife tour operators use green energy to electrify bush cabins and hotel facilities. Climate change and variability are among the current global environmental challenges. The widespread use of biofuels is one of the causes of climate change.

Responsible travelers automatically alter the behaviors of wildlife tour operators by requiring the use of clean, green energy for the electrification of transport and accommodation facilities. As a result, responsible wildlife tour operators have also adopted electric-powered tourist vehicles for tours. Some use solar power to electrify their vehicles and accommodations.

Ethics and Quality in Responsible Wildlife Tourism

Through moral persuasions, ethics are essential in guiding wildlife travelers, wildlife tour operators, or any human towards doing what is right or incorrect, evil, or good actions or decisions. Likes and dislikes express good or right and bad or incorrect values, which, in some cases, become vulnerable to misjudgment because different people may express the adequacy of the same value differently.

However, the expression of an individual's likes and dislikes may vary based on cultural, educational, religious, and political heritage. An excellent value for one individual or group may be a poor value for another. For example, while trophy hunting sounds moral and sustainable to trophy hunters, it seems inhumane and unethical to animal activists.

In that case, consistent judgment standards are required for rational people to judge good from evil and right from wrong. It calls for moral rules, a translation of judgment standards into prohibited and permitted actions, which may reduce the atrocities and increase compliance behaviors, as moral rules may punish non-complying parts and motivate complying parts.

The moral system possesses sets of values, norms, practices, rules, identities, and institutional structures to prohibit or control self-egoism. It involves knowing what actions are morally discouraged and encouraged. However, laws differ from the concept of morality because the law is in a formal public system with written rules, penalties, and judges for interpretation and judgment, indicating the presence of legal authority but not moral authority.

According to the Global Code of Ethics for Tourism, tourism entities must be environmentally, culturally, and socioeconomically responsible. Compliance with international tourism codes reduces the avoidable adverse impacts of wildlife tourism in ecologically and culturally sensitive areas. Like any other stakeholder, wildlife-based tour operators are equally responsible for educating visitors to act responsibly and reasonably.

As companies working closely with wildlife tourists and conservation authorities, it is vital to assist wildlife tourists in following all principles of responsible tourism to save the fragile habitats, species, and communities adjacent to natural habitats. The codes guide sustainable and responsible tourism activities for all entities.

Non-compliant tour operators are likely to exacerbate many forms of wildlife crime, including wildlife trafficking, poaching, habitat devastation, climate change, and other social crimes such as prostitution, narcotics, and human trafficking. Such impacts are expected to occur within wildlife attractions and adjacent communities where frequent visitations by wildlife visitors are common. Thus, the adverse impacts attract legal and moral adherence to principles of sustainable wildlife development from wildlife tourism stakeholders.

The wildlife tourism stakeholders may ask themselves what I am obliged and not obliged to do for their business to be legally or morally responsible. The economic, sociological, and psychological factors usually outline the context, rationality, and structure of their questions.

Several studies investigate the effectiveness of deterrence, or legal and moral persuasion, in the compliance behaviors of wildlife-based tour operators with the principles of responsible tourism. Whether both deterrence and morality jointly work effectively, morality enforcement solemnly functions advantageously, or legal persuasion alone would appear to desirably work effectively in the enforcement of compliance with principles of responsible tourism.

According to some notable unethical or immoral issues in wildlife tourism, the mistreatment of wild animals by tourists and tour operators is at the top. Uncontrolled wildlife tourism causes several unethical issues, such as cage hunting, wildlife trafficking, animal feeding, trophy hunting, and animal disturbances.

The intention behind cage hunting is justifiable, as it considers the involvement of impaired persons in wildlife tourism, but it denies the animal the opportunity to flee, as every pursued animal has a right to escape in trophy hunting.

Also, wildlife tourism may contribute to wildlife trafficking, which in most cases involves unprofessional and illegal capturing and transportation of wildlife species. During capture, the social, ecological, and psychological aspects of captured individuals are intentionally ignored. Also, animal feeding is immoral as it makes untamed animals beggars, and it inherently domesticates the animals by making them incapable of foraging for themselves.

It is evident in the busiest protected areas, as well as those that have been bisected by main highways. Due to the regular and casual feeding of wild animals, you find wild animals queuing up along the roadways, waiting for the remnants tossed out by passersby. In many cases, these animals find it easier to wait and feed on leftovers than to forage themselves. And such animals are typically emaciated and malnourished due to perpetually grazing on harmful food items.

Also, because such animals heavily depend on remnants, it is simple for wildlife traffickers and poachers to capture them with poisoned food products. In the definition of wildlife, it was explicitly stated that untamed animals should be free-ranging without being supplied by anyone.

Feeding the animals denies them ecological and social responsibilities, as the animals will not be able to forage and support their individuals, ending up with malnutrition and ultimately fatalities. Sustainable wildlife tourism ensures that all animals' untamed features are preserved and treated with respect and compassion.

.

Innovations in Responsible Wildlife Tourism

Innovations entail the renewal or improvement of existing wildlife tourism products, services, and operations by implementing new ideas or techniques to create new ways or services of wildlife tourism to heighten tourists' experiences and safeguard the integrity of wildlife resources and local communities.

Innovations in wildlife tourism imply a novel but appropriate method for operating it, a new wildlife product or service, or a new wildlife tourism business model. Innovations in wildlife tourism remain essential for the integrity and maintenance of wildlife resources and the competitiveness of the wildlife tourism enterprise.

What will happen if wildlife tourism ignores innovation? As many innovations in wildlife tourism are directly linked with the policies of sustainable development, wildlife tourism without innovations is hazardous as it may further enhance destructive activities, including poaching incidents, environmental destruction, and social crimes, among others.

Innovation means enhancing wildlife tourism and overcoming any unpleasant challenges associated with it for an improved wildlife tourism experience, wildlife conservation, and local people's livelihoods. In other words, innovations imply the act of wildlife tourism stakeholders taking on board all requirements of responsible wildlife visitors by adopting sustainable wildlife tourism.

In Chap. 6, we will have a comprehensive discussion on sustainability, which means that innovation in wildlife tourism should occur in accordance with sustainable wildlife tourism principles. They should concentrate on sustainability and conservation, just as the destinations should develop and diversify their economies with biodiversity conservation in mind.

Innovations in wildlife tourism should reflect all spheres of wildlife tourism, from destination to attraction levels. All constituents of wildlife tourism, including transportation, accommodations, amenities, attractions, and ancillaries. Each wildlife tourism entity may benefit from constructive innovation, which may improve the character of tourism services and products.

More importantly, technological innovations become inevitable because they may help destinations strengthen their industries, exceed visitors' expectations, and preserve the natural environment. Wildlife tourism should adapt to evolving global technologies. Customization of existing technologies to suit the context of wildlife tourism or innovation of specific technologies to enhance wildlife tourism development is essential.

Robotic and artificial intelligence (AI) technologies are some of the means for enhancing wildlife tourism development. I understand that such adoption has occurred in certain destinations around the developed world. Unmanned vehicles, air taxis, and augmented reality are some of the technological innovations presently implemented in the developed world.

Also, the deployment of robotic technology in the hospitality industry is progressively emerging in developed countries. Despite the technological adoption of state-of-the-art technologies in wildlife management, including satellite and unmanned aircraft for anti-poaching operations, wildlife tourism still requires relevant technological innovation to heighten wildlife tourists' experiences and satisfaction while maintaining acceptable environmental and socio-cultural standards at the destinations.

Relevant technological innovations are critical for the successful transformation and enhancement of wildlife tourism. An integration of augmented reality and artificial intelligence may collectively form automated tour guides to transport visitors around without the assistance of the chauffeur and tour guides. This may reduce superfluous human errors and emotions during wildlife tour expeditions, thereby increasing tourists' experience and satisfaction.

Safety and Security

Wildlife tourism is associated with several hazards that are likely to cause severe injuries, fatalities, and the destruction of properties. In wildlife tourism expeditions, either humans or untamed animals may accidentally perish or acquire serious injuries. Forget about those wildlife tourism activities that involve the intentional slaughter of untamed animals for trophies. For that reason, prioritizing the safety of human and animal weapons is inevitable.

In summation, safety means the protection of wildlife visitors, wildlife resources, wildlife tour operators, and the entire wildlife attraction from various risks, offenses, natural catastrophes, health hazards, and accidents, among others. Any wildlife tourist destination's goal is to have wildlife expeditions with no injury, property damage, or fatality. Safety and security are critical components of high-quality wildlife tourism services, so any destination must have a relevant safety and security policy in place.

Without appropriate safety and security strategies, wildlife tourism jeopardizes the reputations and achievements of various tourism businesses. This suggests that unsafe and precarious wildlife tourist destinations make international tourists vulnerable to terrorists, bandits, and communicable diseases at the destination. Terrorists may target unsecured international visitors arriving from certain countries with doubtful diplomatic relations.

Also, destinations with political, social, and economic instability and conflicts imperil the security and safety of wildlife visitors. It is therefore essential for tourist destinations to secure all tourist entrances and wildlife attractions to safeguard the lives and property of wildlife tourists.

In the same manner, the destinations and wildlife tour operators should inform prospective wildlife tourists in advance of the probable communicable diseases and appropriate vaccinations to execute before and during their travels. Wildlife tourists can be vulnerable to communicable diseases due to a lack of natural immunity.

In that case, it is important to create a precautionary awareness campaign for wildlife tourists. Safety and security may influence wildlife tourists' decisions about the selection of tourist destinations, their satisfaction with wildlife tourism products and services, destination exploration and utilization, and their behaviors. It is evidently difficult for many wildlife tourists to explore the entire destination when there is no serenity, safety, or security, though some gloomy tourists would like those situations.

It is impracticable for wildlife visitors to engage in outdoor adventure activities without any relevant safety, security, or rescue measures. It is evident that adventure wildlife visitors take calculated risks. On that note, without adequate security and safety measures in place, even adventure travelers may disregard hazardous destinations or adventurous activities. Prioritizing safety and security at the destination is key to successful wildlife tourism development. The success and failure of wildlife tourism depend on political, social, and economic stability.

Remember that the government and all stakeholders are responsible for destination safety and security. Since every stakeholder has a unique role to play in wildlife tourism development, it is mandatory for each of them to routinely enforce safety and security measures on their part.

Several factors may contribute to unsafe and insecure wildlife tourism operations, including political, economic, and social instability; geographical location of the attraction; environmental conditions; tourism infrastructure and superstructure characteristics; safety and security policies at the destination; and preparedness of the wildlife tourists.

Political, economic, and social instability may directly or indirectly affect the safety and security of the destination. Social and political instability makes the destination unsafe and insecure because the entire destination usually becomes agitated, disengaged, and inaccessible with limited social transactions due to the absence of norms and laws.

As a result, there is an increase in armed robbery, rape, and terrorism incidents at the destination. In certain circumstances, some members of society may be reluctant to participate in wildlife activities because they are not part of wildlife tourism activities.

Additionally, economic instability may adversely affect the safety and security of wildlife visitors by implicating financial transactions and tourism businesses at tourist destinations. Economic instability amplifies business competition among wildlife tourism stakeholders, necessitating some businesses to adopt unethical tactics to maximize their profits.

As a consequence of economic instability, many wildlife tourism stakeholders fail to fairly compete in wildlife tourism enterprises. Such tactics also undermine the safety and security of wildlife tourism by offering substandard products or services. Imagine the situation when the wildlife safari operator engages in illicit drug businesses or money laundering in collaboration with international travelers to maximize their profits.

Moreover, try to imagine the safety risks from the situation when travel operators, including the owners of airlines, ships, or tourist vans, buy cheap or counterfeit spare parts to fix their equipment or hire unqualified and cheap pilots to transfer wildlife tourists just because they want to maximize their profits.

As described, the destinations are characterized by countless accidents due to inadequately maintained and operated equipment. The geographical configuration of the destinations or wildlife attractions may affect the security and welfare of wildlife visitors and wildlife tourism activities. Certain wildlife destinations are situated in earthquake-prone areas, mosquito-contaminated areas, severe climates, and difficult terrain, which make them not only vulnerable and hazardous but also difficult to access.

One could argue that wildlife attractions' challenging geography is what makes them unique, enjoyable, and memorable. I don't believe the statement may equally apply to all wildlife tourists. In fact, it depends on the type of tourist, budget, age, objective of visitors, and the extent to which the tourists can take risks.

Given the above, what is the purpose of appreciating the inhospitable geographical settings of the attractions while ending up as a deceased or severely injured wildlife tourist? Mountains are some of the attractions with challenging geographical locations. Did you know that every year, 2455 mountaineering tourists die from ascending mountains? Aquatic destinations are similar.

With that said, to a certain extent, I agree that a challenging geographical setting catalyzes the uniqueness of the attraction and destination. However, it is also essential to highlight that well-regulated and monitored wildlife tourism activities in challenging geographical attractions or destinations are vital for avoiding preventable fatalities, injuries, and loss of property during wildlife touring expeditions.

The condition of tourism infrastructure and facilities affects the security and safety of wildlife stakeholders. The geographical location, construction materials, design, and technology may affect both the tourism infrastructure and the superstructure.

Substandard transport infrastructure and facilities built in earthquake-, inundation-, and debris-sliding-prone areas jeopardize the lives and property of wildlife tourists. To avoid unnecessary accidents, it is essential that tourist infrastructure and facilities such as airports, roads, transport facilities, and accommodation facilities be improved.

Formulation and adoption of safety standards for every wildlife stakeholder to prevent and manage risks. Effective safety and security policies at the destination or attraction level may assist wildlife stakeholders in selecting construction sites, construction materials, designs, and technologies.

Moreover, the adoption of reputable international safety and security standards may also help save the lives and properties of wildlife tourism stakeholders. The characteristics of wildlife travelers usually affect the safety and security of stakeholders by exposing them to unfamiliar environments and activities.

Such characteristics usually vary based on the gender, origin, culture, and education of wildlife tourists. Some wildlife visitors may take unnecessary risks by interacting with an unfamiliar culture or environment to satisfy their requirements.

Wildlife tourists are usually informed in advance of the probable risks of indulging in hazardous tourism activities, including mountain climbing, scuba diving, snorkeling, and trophy hunting. The desire to fulfill their life ambitions makes the minds of wildlife tourists unchangeable, despite their existing health and other conditions. As a result, wildlife travelers engage in activities that not only ruin their lives but also damage their properties.

Several incidents exist in protected areas with large surface areas where wildlife travelers conduct game driving without a tour guide and driver but end up deceased in the car after losing direction. Moreover, the preparedness of the wildlife visitors before and during the wildlife tourism expedition is essential, as it creates awareness of the potential risks and effective mitigation measures.

A timely awareness creation for wildlife tourists on the existing environment, culture, and economic risks at the destination may help wildlife tourists decide on the selection and visitation of hazardous areas.

Satisfaction of Responsible Wildlife Tourists

Satisfaction in wildlife tourism entails meeting or exceeding the expectations and preferences of wildlife travelers participating in wildlife tourism without contradicting the environment and socio-cultural aspects of the destination. Several factors determine the contentment of wildlife visitors, including service, information, accessibility, and the value for money of tourism products and services.

Alignment of the qualities of wildlife tourism products and services to local and international standards is one of the methods of meeting or exceeding the expectations of wildlife tourists. Wildlife tourism is a business, just like any other. To satisfy customers' needs, the resultant products and services must meet or exceed their expectations and preferences. To attract a large number of wildlife visitors, the services and products from wildlife tourism should be relevant and of high quality.

Given the above, several factors may compromise the satisfaction and preferences of wildlife visitors. Apart from the quality of wildlife products and services enumerated above, accurate information, value for money, safety, and security, and accessibility of the services and products may severely affect the satisfaction and preference of wildlife visitors. A lack of sufficient information about specific wildlife tourism products and services may affect wildlife visitors' satisfaction and preferences.

Accurate descriptions of every wildlife product or service offered at the destination may help wildlife tourists determine, accordingly, which wildlife tourism services or products to consume. Also, the accuracy and logic of the information provided during the wildlife tourism expeditions are important for meeting or exceeding the preferences or expectations of wildlife tourists.

This suggests that wildlife tourism requires knowledgeable, experienced, and competent tour operators and administrators to provide an infinite experience to wildlife tourists. Furthermore, safety is an essential ingredient in providing quality tourism services and products. Unsafe destinations and attractions can hardly offer quality tourism services and products.

It is hard for wildlife tourists to utilize unsafe tourism services, products, or quality wildlife services in an unsafe destination. Moreover, inaccessible quality tourism products or services are as substandard as other products and services. Wildlife tourism should make sure that all products and services are accessible.

Chapter 8
Challenges in Sustainable Wildlife Tourism

Abstract This chapter emphasizes the challenges facing wildlife tourism from local to global contexts. Just like other forms of tourism, wildlife tourism is not free from obstacles. Some of the constant challenges facing wildlife tourism include engagement among communities, global crises, conserving biodiversity in the era of commercial poaching and human–wildlife conflicts, and climate change and variability. Insufficient local and international cooperation from various wildlife tourism stakeholders is one of the challenges that make wildlife tourism fail to balance between ecological integrity, social equality, and economic viability, which is essential in achieving sustainable and responsible tourism in destinations with limited resources. The concept of sustainable wildlife tourism can be challenging to execute, even with the aid of open sustainability certification programs, suitable technologies, laws, and regulations that support without proper support from tourism stakeholders. Wildlife destinations experience many challenges during species conservation; mitigating human-wildlife conflicts is still challenging because of the increasing human population and climate change. Insufficient sectoral and international technical, logistical, and financial support for addressing the negative impacts of commercial poaching and wildlife trafficking are anticipated to adversely affect wildlife tourism in the near future. Climate change and unpredictability challenge wildlife tourism due to their continual harmful impacts on wildlife habitats and behaviors. Climate change through extreme rainfall and cyclones may harm tourist infrastructure and superstructure in particular destinations. By altering ecosystems and tourism infrastructure, climate change makes the execution of sustainable principles even more difficult and amplifies the costs for infrastructural recovery and adaptive management. As one of the global crises, the COVID-19 outbreak demonstrates the way wildlife tourism could be affected by the global crisis. During the COVID-19 era, wildlife tourism took place on a limited scale and cautiously to prevent the future transmission of the diseases. This underscores the requirements of employing virtual tourism technologies and the adoption of market diversification techniques to promote sustainable and responsible wildlife tourism. Thailand and Bhutan exemplify the ways conservation, socioeconomic growth, and cultural preservation may be linked and integrated into sustainable tourism frameworks to attain the high-value, low-impact tourist model.

A. A. Mamboleo, *Sustainable Wildlife Tourism*,
https://doi.org/10.1007/978-3-031-97795-4_8

Keywords Community engagement · Climate change adaptation · Human-wildlife conflict · Eco-friendly tourism strategies · Tourism revenue and conservation · Stakeholder collaboration

To overcome such difficulties, wildlife tourism stakeholders need to adopt outstanding abilities, knowledge, and abilities. Without solving such problems, the future of wildlife tourism is always uncertain. As a result, wildlife tourism stakeholders frequently collaborate to address some of the issues with wildlife tourism. It is challenging for a single stakeholder to overcome such hurdles.

Thinking outside the box or having helicopter skills are additional talents for overcoming obstacles. Most issues of wildlife tourism are comparable to those of wildlife conservation since wildlife resources are the major assets of wildlife tourism.

Without wildlife resources, wildlife tourism would live in vain—that is to say, no wildlife, no wildlife tourism. The foregoing sentence does not imply any wildlife or no tourism; instead, it implies no wildlife and no tourism. What about no wildlife tourism, no wildlife? The simple answer to the above assertion is that wildlife can continue to live even in the absence of wildlife tourists.

Even though wildlife tourism provides various benefits to wildlife conservation, it is not the primary and only cause of wildlife existence. However, while the discussion on wildlife and wildlife tourism may be fierce and ongoing, the crucial thing to take away is that wildlife tourism and wildlife conservation sail in the same boat by sharing comparable issues, most of which are technically created by people. Also, some of the issues experienced by wildlife tourism locations in underdeveloped countries are comparable.

The problems of wildlife tourism, among others, may include the ability to protect places with maximum biodiversity and species endangerment. Get locals involved in wildlife tourism, make more foreign currency from wildlife tourism, connect wildlife tourism with household development, make wildlife tourism more competitive compared to other types of tourism, reduce the negative effects of wildlife tourism on wildlife resources and communities, do more scientific research on wildlife tourism, and help local investors by raising awareness and creating the right conditions.

Conservation of Biological Diversity

The preceding paragraph aimed to identify numerous issues affecting wildlife tourism, some of which are beyond the capabilities of a single stakeholder. The first mentioned problem, e.g., is that the protection of biodiversity hotspots is expensive because it always requires a large investment of financial and non-financial resources in the whole process.

As the major goal of many protected areas, conservation organizations normally prioritize by dedicating most of their resources to the protection of biodiversity. Biological biodiversity is a fairly broad term, but it refers to having numerous kinds or species of wild animals and wild plants in a single region. It also denotes having multiple species, with many individuals of each species in one place.

The site where biodiversity emerges might be as simple as protected or unprotected terrain. It may also be a nation or area. Animal species are not uniformly distributed globally, regionally, and locally; as a result, some places have more wildlife species than others, and other areas have more individuals of specific species than others. Several causes may explain such different distributions; some of the explanations may include climate, human pressure, and the availability of biological resources.

Additionally, certain locations are worldwide known as biodiversity hotspots due to their high species variety and abundance. An example from the East African area is the Eastern Arc Mountains, which are part of the worldwide biodiversity hotspots. Technically, biological hotspots are places endowed with numerous wildlife species, of which the majority are indigenous, endangered, and unusual.

Such locations may have a variety of wild flora and wild animals, which are both prolific and fascinating to wildlife visitors. One of the most pressing conservation issues is the protection of biodiversity and biodiversity hotspots. With limited and uncertain financial and nonfinancial resources, protecting biodiversity hotspots has become a technical challenge for conservation organizations and governments.

As a challenge, conservationists must think outside the box by asking for external financial and non-financial assistance to rescue biodiversity hotspots in the blink of an eye, which, in this context, are also the key assets of wildlife tourism.

It is vital to emphasize that most of the biodiversity hotspots are situated in developing nations, whose budgets are inadequate even to satisfy two-thirds of the demands of all residents. That is to say, it is impossible for poor nations to fully support conservation efforts without external aid or to prioritize biodiversity protection over human necessities.

Various wildlife stakeholders generally provide the conservation agency with hands to overcome the obstacles of protecting biodiversity hotspots. The aid might be in the form of information, skills, equipment, and funding. As a result, local and international organizations have been at the forefront of solving the problem of biological hotspot extinctions by voluntarily providing financial and non-financial resources to conservation agencies.

Stakeholders participating in biodiversity conservation initiatives may either be solely wildlife tourism operators, wildlife conservation foundations, or wildlife tourists. It is typical in undeveloped nations to see organizations such as the WWF, IUCN, USAID, AWF, and FZS providing funding or helping technically overcome obstacles to biodiversity conservation.

Similarly, wildlife tourism operator enterprises do equally well to help conservation organizations solve the difficulties of biodiversity conservation by providing financial and non-financial logistics. However, despite the passionate and voluntary engagement of wildlife tourism stakeholders in conquering biodiversity conservation, the problem may remain, particularly when the components of political will and community involvement are absent.

It is constantly in mind since Prof. Alexander Songorwa, in one of his lectures, remarked that wildlife managers do not manage wildlife but people's attitudes and views towards wildlife. In other words, Professor Songorwa technically means that wildlife is not like livestock, which requires the livestock keeper to spend his entire time looking at the livestock, but wildlife managers or conservationists have to think outside the boundaries of the protected areas and understand what people usually want or think of wildlife.

In reality, it is a simple truth: conservation agencies may have all the resources necessary on earth, but without voluntary support from the political sphere and communities, biodiversity conservation could remain as tough as it always has been. It is unfeasible to preserve or even discuss biodiversity without community engagement, especially in underdeveloped nations where biodiversity conservation comes at the expense of local populations.

It is biodiversity that supplies daily meals to local populations without proper consideration of their efforts or the priority of their needs. Conquering biodiversity conservation issues may be virtually impossible.

Participation and Involvement of Residents

Those who frequently and involuntarily suffer most of the conservation costs in terms of human and livestock mortality from problem animals, crop loss, and indirect expenses. Also, as individuals devoting their lands to conservation while founding, enlarging, or improving the protected zones, local people should equally and equitably be recognized in wildlife resource exploitation via wildlife tourism.

Local people's direct and indirect involvement in wildlife tourism operations is a historical problem for certain administrations. As one of the strategies to tackle it, specific concerns have been taken on board by various governments by actively engaging the local community directly. With the community-based natural resource management paradigm, some governments legally empower the local community to maintain and exploit wildlife resources in their village territory for the benefit of the communities.

On that basis, wild flora, wild animals, and wild areas in the village land are sustainably maintained and exploited by the people under the direction of the governments. The strategy stated above is important as it directly incorporates local populations into the wildlife tourism industry via consumptive and non-consumptive wildlife resources. It effectively addresses the problem of incorporating people into wildlife tourism.

However, it also still confronts some obstacles because, if not cautiously implemented and controlled, decentralization of wildlife management and utilization to local people may become a burden. Instead of rewarding locals, it may put them in trouble. In developing nations, illiteracy and poverty rates in rural regions are relatively high; these are the same locations where local groups are permitted to manage and exploit animal resources.

Wildlife tourism involves local and international business enterprises with more sophisticated business skills and experience than local communities; authorizing local people to manage and utilize wildlife resources without any prior legal, business, or conservation knowledge sounds awkward. The point I was trying to make in the preceding paragraph is that enabling the locals to manage the protected areas is one thing, and being able to manage and exploit the animal resources legally and professionally is another.

Close technical and financial assistance from the government is needed to completely empower and appropriately include local populations in the wildlife tourism sector. It is safe to say that I have now and then been cited throughout this book that conservation is usually expensive, and only a few countries with viable wildlife resources can afford to fully fund it.

Just try to imagine: if the government in the developing countries fails to completely fund and manage all protected areas for 100 percent, can the poor and highly indebted peasants in the local communities afford to fund their protected landscape for at least 10%?

Surprisingly, some community-based wildlife reserves have larger surface areas than public ones; hence, they need more resources and technical skills to administer. In that setting, the direct involvement of local populations in wildlife tourism becomes complex and demanding, particularly in impoverished countries. It consequently seeks external technical and logistical aid from interested wildlife tourism players to empower the local community.

The attainment of external support is not as straightforward as it appears, since diverse potential supporters may have different interests and criteria. Furthermore, some sponsors may be interested in supporting specific animal species, locations, types of protected areas, or natural elements such as rivers, lakes, or the ocean. Other sponsors, instances, and research and education institutes may be interested in assisting the local community's technical knowledge and abilities, either through training or research.

The supply of government subsidies and technical support is the next approach to helping local communities manage wildlife resources successfully, where governments may regularly provide financial and non-financial support. Also, the various means of incorporating local people in wildlife tourism include indirectly helping local communities' development measures. Cost-benefit sharing from wildlife tourist activities also qualifies as indirect engagement of local people in wildlife tourism.

In this context, the governments remain with the main responsibilities of managing the protected areas, while locals indirectly participate in conservation as good Samaritans by alerting the conservation authority on the occurrence of any wildlife crime, but sometimes local people lose their lives and property for the sake of conservation. All these are termed conservation costs. Similarly, some governments redistribute a certain percentage of the profits from wildlife tourism to local people in a variety of ways.

The proportion given to the communities may vary based on government policy; for example, the Tanzania Wildlife Management Authority (TAWA) returns 25% to the community annually. The same applies to wildlife tour providers that also return a specified percentage of the advantages derived from wildlife tourism to the community. The most prevalent approach is the construction, maintenance, and rehabilitation of social infrastructure.

The governments develop hospitals, schools, clinics, and roads with the money they earn from wildlife tourism. Communities around wildlife regions are the primary benefactors and victims of wildlife tourism. In most situations, the villages are contacted in advance to recommend the sort of social projects or infrastructure they would like to have created.

Additionally, the protected areas, directly and indirectly, generate official and informal job possibilities for locals via the sale of local food products and souvenirs to wildlife visitors and housing facilities. Some protected areas employ locals as tour guides and rangers. Also, the existence of protected areas in their proximity offers the residents commercial possibilities by selling different local items to visitors and hotels as a method of raising family earnings.

Balancing Revenue and Conservation

Even though it still faces challenges in increasing its foreign currency earnings, wildlife tourism is one of the major sources of foreign money in developing nations. According to FMI, the worldwide wildlife tourism industry is expected to be valued at US$135 billion in 2022 and is projected to expand by 5% from 2022 to 2032. According to FMI, worldwide wildlife tourism generates an enormous amount of foreign currency, totaling $232.6 billion USD, which is comparable to Hong Kong's entire economy.

The figure from FMI indicates the global earnings of wildlife tourism, but when the figure is divided by the number of countries that offer wildlife tourism, the amount may not be enough. Also, considering the financial and non-financial resources invested in wildlife tourism development, the foreign earnings still look like peanuts. According to WTTC, Asia-Pacific is the largest regional market for wildlife tourism, worth US$ 53.3 billion in direct GDP and 4.5 million in employment prospects.

Africa's continent comes in second, with 3.6 million people employed by wildlife tourism, which was valued at US $29.3 billion in 2018. According to figures published by WTTC, wildlife tourism serves as the backbone of tourist operations in many African nations because it directly adds to GDP and produces official and informal employment, although its real impact on foreign revenues is still uncertain. Africa's continent comes in second, with 3.6 million people employed by wildlife tourism, which was valued at US $29.3 billion in 2018.

According to figures supplied by WTTC, wildlife tourism provides the backbone of tourist operations and the major source of foreign cash in many African nations since it directly adds to GDP and produces official and informal employment. The yearly foreign currency revenues from wildlife tourism are unappealing; numerous causes may account for this, including inadequate recording, incorrect figures, and a lack of transparency in reporting foreign earnings. It is conceivable for certain governments to boost or lower the amount of foreign profit from wildlife tourism to correspond to their interests.

Similarly, some overseas revenues may be incorrectly documented because of an outdated recording system or dishonest workers. Moreover, documenting other profits might be technically problematic, especially in the enclave locations where most of their payments are made outside the host countries. Furthermore, it is frequently difficult to distinguish between the revenues generated by wildlife tourism and those from other categories of tourism, such as cultural, medical, and religious tourism.

Certain locations or protected regions contain cultural, medicinal, and animal treasures. It is vital to emphasize that foreign revenues from wildlife tourism are significant because they support wildlife destinations and protected areas in tackling financial and non-financial challenges.

It was previously observed that wildlife does not always require wildlife tourism to survive. In certain areas, wildlife tourism is vital for supporting not just conservation operations but also helping the governments fund some of the development activities.

Foreign exchanges allow governments in international enterprises to acquire medical supplies, energy, and other critical international items that could not be produced locally. In brief, wildlife tourism generates an inadequate quantity of foreign cash compared to conservation expenditures. Conservationists and economists should uncover the technique of raising foreign profits from wildlife tourism.

The profits can cover all conservation expenses and help governments fund the majority of social programs. Similarly, governments should establish clear and efficient documentation mechanisms to boost foreign profits from wildlife tourism. Moreover, the government should sensitize other wildlife tourism players to the need for documenting and reporting the revenues obtained from wildlife tourism.

Climate Change and Variability

Climate change is one of the concerns confronting wildlife tourism, as it is likely to wipe out most wildlife tourist activities in the future. It impacts all industries and nations in various ways, directly or indirectly. In that respect, the tourism industry is not exempt from the ghosts of climate change, as it constantly influences wildlife tourism via species extinctions, migrations, and behavioral changes.

As a sort of nature-based tourism, wildlife tourism continues to be one of the most impacted owing to the excessive dependency on animal resources and habitats as key attractions, whose survival depends on sufficient rainfall and stable climatic conditions. Climate change impacts wildlife tourism owing to the vulnerability of wildlife resources to climate change for the prolonging and preservation of vital biological processes.

Wildlife tourism frequently uses wild flora, animals, and their natural surroundings to attract visitors, which is undoubtedly one of its key advantages. Prolonged droughts and cyclones affect wildlife habitats and species, which are the main assets of wildlife-based tourism. Climate change has been affecting the global tourism industry continuously through the destruction and deformation of tourist attractions, infrastructure, tourists, and facilities, which are the backbone of the industry.

Its repercussions, like sea-level rise, continue to endanger the survival of treasured coastal sites and destinations. Several tourist islands in the sea have perished, and many more continue to sink. As Albert pointed out in 2016, at least 11 islands in the northern Solomon Islands have either fully gone or are in the process of submerging and becoming uninhabitable shortly. As a prominent worldwide tourist destination and the country with the lowest terrain in the world, the Maldives will be fully submerged and hence become uninhabitable by 2050.

Lengthy droughts put some places in danger of starvation, while regular floods and cyclones harm tourism infrastructure and services in cyclone-vulnerable tourist destinations such as Thailand, the United States, and India. Wildlife-based tourism as a total experience, from observations to interactions with wild animals and plant life in their naturally occurring habitats, extinctions of wildlife species, and disappearances or destructions of wildlife habitats, may contradict the future of the wildlife tourism industry.

Climate change and variability impact the reproductive habits, population structure, sex ratio, and age structure of animal species. It also influences the migrations and migratory patterns of terrestrial, aquatic, and arboreal animal species, resulting in the devastating local and worldwide extinction of charismatic species. The decline of iconic species and their habitats may lead to either devaluation or the outright extinction of the most prominent wildlife tourist locations.

This may have environmental and socio-economic ramifications for nations and populations that substantially depend on animal resources. Diversity and quality of animal resources influence the strength, potential, and dependability of sustainable wildlife tourism.

Such characteristics may have an even greater impact on the variety of wildlife tourism goods and services available at the site. Because wildlife resources function as the cornerstone of wildlife tourism activities and attractions, the repercussions of climate change restrict the existence of animal species and their ecosystems, making the future of wildlife hazy.

Such consequences may lead to local or worldwide extinctions of some flagship wildlife species; the disappearances of such species may dissuade wildlife visitors from coming and returning to the East African sites. The loss of attractive species and their related ecosystems devalues the destinations, rendering them completely or partially unappealing.

The literature provides infinite cases where the consequences of climate change and variability constantly worsen the future and quality of wildlife tourism. The literature claims that climate change impacts most of the aspects of wildlife tourism, including natural and man-made attractions, transit, and housing amenities.

Balancing Wildlife Tourism and Household Development

It is deceitful to speak about the advantages of wildlife tourism without mentioning its benefits at the household level. In many situations, wildlife stakeholders perceive the advantages of wildlife tourism as social infrastructure and services provided to communities at the organizational level.

At the household level, it is easier to assess the direct consequences of wildlife tourism for households and its contribution to poverty reduction via economic empowerment. Frankly speaking, the integration of the advantages of wildlife tourism has been a difficulty for wildlife tourism for years.

Everyone understands the conservation costs households incur owing to wildlife, which is the main asset for wildlife tourism. This is the household unit that loses crops and other possessions to wildlife. It is the same home whose heads and members die or incur significant injuries from animals.

Some members of families are wrongly imprisoned or tortured for animal offenses. The village is only an administrative entity that manages houses, not land or people. It can be difficult to feel the ache of losing a household member or possessions.

The village government cannot detect when family members starve. The village government's functions are to provide a peaceful environment and social services at the local level. With all those being stated, it is time for wildlife tourism stakeholders to explore the collective advantages of wildlife tourism with household development. At the moment, integrating the advantages into home socio-economic growth appears to be difficult, both technically and monetarily.

The mechanism for benefit distribution can be another challenge to overcome, as it is still unknown if all household units in a particular wildlife tourist destination or country should equally get the benefits from wildlife tourism or if the benefits should only be distributed to those households neighboring the protected areas. The households near the protected areas continually and immediately bear the conservation expenses, which scarcely make local people successful big-scale commercial agriculturists, making them impoverished and inferior.

Prioritizing such homes may be fairer and more comforting than the present consolation plans, which mandate certain governments to console afflicted households by paying out a set amount of money following property or life loss from households due to animals.

It is crucial to have a unique method and mechanism in place for the distribution of wildlife benefits down to the home level. With that in mind, the transfer of cash or hard money to homes may not benefit the households; instead, the government should encourage the households to launch their revenue-producing enterprises for household development.

Alternatively, the government may offer an acceptable quantity of loans with zero interest for the families to participate in a feasible enterprise. However, frequent training on business loans and repayments to heads of family units is critical for their enterprises' sustainability. Apart from loan provisions, the government may disburse the advantages of wildlife tourism to families by granting subsidies to some of the basic commodities or products sold in the communities next to protected areas.

Since agricultural operations tend to fail in such locations owing to frequent animal incursions, the governments waive or offer free medical and education services to all families. Additionally, supplies of discounts on food products sold in certain settlements may stimulate and let the inhabitants experience the advantages of wildlife tourism. However, the above-proposed ways are often difficult to adopt because of the country's distinct natural resources and economic policies, as well as the accompanying negative consequences.

Such techniques may later promote and entice many families from distant communities to relocate to such regions aimed at reaping direct advantages from wildlife tourism, since locations near protected areas may have comparably superior social services than other areas. As a result of increased living standards, overcrowding will be the case in such communities, leading to environmental and socioeconomic implications in the villages next to protected areas.

All things typically cause adverse impacts to conservation, since incidents such as wildlife crimes may also grow. Population growth harms wildlife, tourists, and residents. As heightened rates of social crimes may equally harm both wildlife visitors and inhabitants,

In such circumstances, stakeholders in wildlife tourism should carefully prepare in advance when thinking about incorporating the advantages of wildlife tourism into home development.

Furthermore, incorporating the advantages of wildlife tourism into household development and enabling members of households to experience the touch of wildlife tourism needs solid clarification on the sources of finances and regions of implementation. As a matter of clarity, it is necessary to highlight when seeking to solve the difficulty of integrating the advantages of wildlife tourism into family development because wildlife benefits may be both financial and non-financial.

Additionally, governments and non-government entities may benefit. Moreover, the advantages might arise from admission fees, concessions, or contributions, where wildlife tourism players may provide technical or other resources to the families who live near protected areas as a method of supporting or encouraging their efforts towards wildlife conservation. However, the government must substantially coordinate contributors because uncoordinated gifts may occur at random and confuse the recipients.

More tragically, certain contributions may arise and operate against the desired intent. It is typical for some of us to think of the advantages wildlife tourism may provide solely in terms of the financial aspect, but such benefits might also extend to non-financial elements.

Offering foreign and local scholarships to well-performing students, sponsoring international or local medical references, building high-efficiency stoves, providing legal assistance to households' units, and training wilderness survival skills in children are some of the non-financial benefits of wildlife tourism.

In such circumstances, a solid understanding of both the financial and non-financial advantages of wildlife tourism is necessary. It assists relevant authorities with the successful planning and distribution of benefits to recipients. It is the responsibility of both governmental and non-governmental groups to provide non-financial resources as an advantage of wildlife tourism.

Furthermore, the non-financial resources may include the direct employment that members of households obtain from wildlife tour operators and protected areas, as well as the opportunity that households have to conduct business with tourist hotels and wildlife visitors.

Competitive Advantage of Wildlife Tourism

Sustaining and improving the competitive edge of wildlife tourism over other types of tourism has become difficult. Wildlife tourism has been regarded as one of the oldest and most prevalent forms of tourist activity internationally. For decades, wildlife tourism activities, notably game viewing and hunting, have been at the heart of the tourism sector. However, sustaining its competitive edge is a major task.

Flexibility and a multi-sectoral approach are some of the important components necessary for wildlife tourism to preserve its competitive advantages over other emerging kinds of tourism. Cultural, recreational, and religious activities are examples of fast-developing activities competing with wildlife tourism.

Considering the fragility of wildlife resources, wildlife tourism seems more fragile than other types of tourism since its existence is sensitive to anthropogenic and non-anthropic activity. The fragility of wildlife resources makes the future of wildlife tourism appear doubtful since it is the aggregation of numerous human and non-human variables that determine its existence.

In other words, it requires destination managers and governments to collaboratively manage and coordinate all anthropogenic and non-anthropic factors to enhance the quality and quantity of wildlife resources for wildlife tourism and maintain its competitive advantage over other forms of tourism. It is difficult because animal resource management is not straightforward and usually has specific financial, political, social, and technological ramifications.

Another problem is managing and balancing all of these domains, which likewise require money, time, and effort. Particularly in public protected zones, where everything in such regions is perfectly done according to established bureaucratic structures, in the process of complying with these systems, some of the unique animal resources may regrettably vanish. Also, wildlife tourism is prone to wildlife crimes, including wildlife poaching and trafficking, which entail the taking of economically desirable animal species for live commerce or trophies.

It is a huge danger to wildlife tourism since it entails the taking of wildlife resources without any concern for their immediate and long-term ecological and social worth in the particular protected area. As a result, many poached or traded species may experience local or worldwide extinction. Ecologically, if wildlife trafficking and poaching target keystone or umbrella wildlife species, their extinction may expand the issue further by equally hurting other wildlife species.

A pretty easy example is the elephant, the animal that most of us have come across either by reading about it in books, seeing it in the media, or seeing it practically. The elephant has been given titles because of the positive ecological responsibilities it performs. As an ecosystem architect, it develops paths in thick vegetation for other animals to pass; it converts from forest to different kinds of vegetation, enabling other species to feed comfortably.

It is referred to as a mega-gardener because it contributes to the dispersion of seeds for more than 300 types of tropical wild plants. Moreover, as an ecosystem engineer, I excavate or dig water holes in dried rivers during dry seasons for myself and other creatures to drink. As bush physicians, elephants are notorious for targeting and clawing particular rocks for themselves and other animals to lick as therapy and mineral supplements.

Similarly, people use elephant dung as a source of medication, notably in Africa, where the dung is locally used as a therapy against cerebral malaria in children under 5 years old. Furthermore, the weak digestive system of elephants allows other wild creatures, especially wild birds, to graze on undigested food from their excrement. Given all the examples and narrations of what elephants can achieve from ecological and social perspectives, failing to sustain such keystone species is likely to damage many other species.

Conservation of keystone species remains a problem in countries with active elephant ranges because of their economic, social, and ecological importance. As a worldwide conservation and wildlife tourism problem, again, the united efforts may be a way to protect the survival of such species. As a result, the complete disappearance of such species may extend to environmental aspects, resulting in the loss of scenic qualities or attractive features of the protected regions.

Because numerous components in the protected regions rely on each other, that is to say, wild plants rely on wild animals, and the converse is correct. Wild plants are sources of food for wild animals; however, the reproductive success of certain plants relies on animals as agents of seed dissemination and pollination. Also, wild animals and plants rely on non-living objects to eat, drink, and procreate. However, non-living aspects, such as soil formation and rainfall, rely on wild plants.

Wild plants exchange gases by releasing water into the atmosphere via transpiration, one of the biological processes that leads to cloud formation. All those are key assets for wildlife tourism; the absence of one may affect others. Some of the off-seized resources in the corporate sector may be unlawfully used to populate local and worldwide protected areas, as well as zoological and botanical gardens.

In nations where desirable species are lacking, private entities may unlawfully involve themselves in wildlife trafficking and poaching to complement their tourist industry with rare species. In the long run, such entities may remain the only areas on the globe with appealing species following the departure of such species from the wilderness. One of the live examples is the black rhinoceros, which became extinct in several publicly protected areas due to poaching and trafficking.

Furthermore, in the long term, other kinds of tourism may outcompete total wildlife tourism owing to heightened costs of initial investment, capital, and upkeep, which are greater in wildlife tourism than in other forms of tourism. Public protected areas, where most wildlife tourist activities take place, need more financial and non-financial resources to construct, maintain, and administer than those required for cultural, medical, sports, and religious tourism.

The protected regions need daily law enforcement operations, research, community involvement, and periodic infrastructure construction and maintenance. Additionally, most public protected areas are conservation-focused, not business- or tourist-focused; they are for public protected areas and other types of tourism.

In public protected areas, aside from considering wildlife tourism as a cash-producing activity, conservation authorities see it as a secondary activity, and the protected areas may still function without wildlife tourism operations.

As the major places where most wildlife tourist activities take place, neglecting wildlife tourism as an essential activity in protected areas may contribute to low-development wildlife tourism. Also, the failure of wildlife tourism to adjust with time may outcompete the wildlife tourism industry.

Wildlife tourism, the earliest type, developed over many periods. One of the fatal phases that wildlife tourism went through was mass tourism when wildlife tourism was largely defined by the massification of irresponsible visitors. In the current day, most of the various kinds of tourism have accepted the alternative notions of tourism, which urge responsible travelers and responsible operators to take part in tourism.

Eradication of Adverse Impacts of Wildlife Tourism

Reduction or reduction of detrimental consequences coming from wildlife tourism confronts conservation authorities and destination management. Conventionally, everyone believes that conservation organizations and destination management are the only ones responsible for mitigating, reducing, and eliminating the negative consequences of wildlife tourism.

It is the role of every tourism stakeholder to proactively minimize any possible problems emerging from wildlife tourism to preserve the environment, culture, the well-being of wildlife travelers, and the socio-economic growth of the wildlife destination. Stakeholders in wildlife tourism may easily mitigate the bulk of the damage through coordinated efforts.

Leaving all duties to a few or one tourist stakeholders, notably the government, may eventually amplify and overspread the unfavorable consequences and significantly harm the environment, local communities, and socio-economic development from the attraction to destination levels.

Wildlife tourism has a blanket negative influence that may even harm regions beyond the site. In such a scenario, one person's or organization's mitigation of wildlife tourism's consequences becomes technically and financially impossible. As a result, many wildlife tourism players are required to participate.

Moreover, minimizing the impacts of wildlife tourism requires destinations to adopt various strategies and policies, including careful planning, promoting local people's involvement, adopting sustainability practices, educating tourists, regular monitoring of wildlife tourism operations, and specifying limits of acceptable use and change.

Most of the measures described, including principles of sustainability, limits of acceptable change and usage, and promotion of knowledge among visitors, are the most significant aspects of sustainable wildlife tourism. In that regard, the techniques have been fully explained in Chap. 6.

Conducting Research in Wildlife Tourism

Tourism research is crucial for gathering and communicating scientific knowledge about the advantages, effects, possibilities, and development of wildlife tourism operations at the attraction and destination levels. Without timely and accurate information, wildlife tourism management may become both challenging and hazardous. Most management choices for wildlife tourism attractions and destinations require trial and error.

However, conducting wildlife tourism is tough, particularly in poor countries characterized by low budgets, inadequate human resources, and a lack of focus on areas of study. Wildlife studies are also difficult, particularly when asking for suitable respondents, obtaining quality data, data administration, and analysis.

Study design is one of the hardest topics in wildlife tourism studies because it depends on which places the researcher wishes to gather data from. A suitable study design helps the researchers capture significant factors affecting wildlife tourism quality and satisfaction.

Research design may entail identifying the correct respondents, demographics, sample size, procedures for data collection, and acceptable language(s) for the survey. Choosing the correct wildlife study design relies on the availability of suitable, skilled, and competent wildlife tourism researchers. Furthermore, designing the population and field of study relies on the researcher's previous experience and expertise.

Wildlife tourism attracts local and foreign visitors, whose mediums of communication differ greatly. It is only a bilingual researcher who can effectively speak with all wildlife tourists or construct brief questionnaires that can be answered by all wildlife visitors.

In that case, the intricacy of wildlife tourism demands researchers with varied professional backgrounds, including hospitality, leisure, community development, tourism, conservation, safety and security, and planning, among others.

In conclusion, creating wildlife tourism research is challenging since it takes numerous talents and expertise from many fields. Consequently, researchers with limited experience undertake wildlife tourism studies in specific places. It sounds immoral since the researcher may end up contradicting the validity, reliability, correctness, and interpretation of the acquired data.

It is a typical practice for an inept wildlife tourism researcher to do studies in places, especially in wildlife tourist locations, with inadequate expertise and little research money. In such settings, wildlife tourism researchers utilize their collected research experiences and talents earned over many years.

Moreover, wildlife tourism studies may encounter obstacles while obtaining data from prospective respondents. The respondents in wildlife tourism research include local people, wildlife managers, destination managers, wildlife visitors, tour operators, and political leaders, among others. Before conducting wildlife tourism research, researchers may need to consider the respondents' availability during surveys, their willingness to participate, and the permission of appropriate authorities for human and animal ethics.

All these demand researchers with outstanding degrees of patience and understanding. Many wildlife tourism projects, particularly those containing human and animal components, require researchers to obtain and gain ethical approvals from human and animal ethics committees in advance. Additionally, researchers need to get permission from different government research agencies and commercial wildlife tourism companies.

Researchers may find research permissions and certificates expensive, lengthy, and bureaucratic. Access to the research sites may be difficult because the terrain of certain wildlife tourism attractions is rough. That causes data collection from faraway places and attractions to be incredibly difficult and expensive. Unfriendly geographical positions and transport infrastructure impede the accessibility of isolated attractions.

As a result, several features and locations of wildlife tourism remain understudied or not researched at all. Apart from collecting quality data, the administration of quality data is another issue for wildlife tourism researchers. Poor knowledge and skills, coupled with a lack of relevant technology, may contribute to poor data quality management. As a result, irrelevant inferences and conclusions were drawn from the acquired data.

Literature indicates that badly handled data may technically contradict the validity, correctness, and reliability of the obtained data. It is crucial for wildlife researchers to carefully maintain the quality of the obtained data to decrease or resolve any existing sampling bias, measurement mistakes, and statistical difficulties. As it was indicated in previous paragraphs, wildlife tourism studies are challenging.

Researchers studying wildlife tourism must consistently manage the intricacies, variety, and dynamism of the tourist sector. Wildlife tourism studies have to technically handle the opposing interests and viewpoints of many tourist actors. Furthermore, the use of the acquired data for wildlife tourism development becomes a barrier because certain locations do not trust researchers or conclusions, particularly those challenging the current leadership or political regimes. It demoralizes academics, research institutes, and funding organizations because some beneficial findings end up stuck on shelves without being executed.

Supporting Local Tourism Investments

Traditional wildlife tourism does not effectively empower and assist local tourist investors. Foreign investors are given more importance and priority than domestic investors. Lack of financial resources is the primary reason why local investors are excluded from wildlife tourism initiatives. Interestingly, several wildlife tourism areas exhibit a preference for foreign investors over local ones, even in investment opportunities that are within the financial means of locals.

Such preferences may be influenced by multiple variables. In addition to inadequate financial resources, political instability, the absence of a supporting investment strategy, a lack of information about wildlife investments, and issues with bureaucracy and corruption may also impede local wildlife tourist ventures.

The aforementioned concerns, whether considered separately or in combination, impose limitations on wildlife tourism destinations in terms of their ability to assist and empower indigenous wildlife investors.

Limited fiscal resources in emerging economies pose challenges for destination countries in providing enough assistance and empowerment to local investors. In economies of this kind, the available funds are consistently insufficient to adequately sustain all sectors.

That is to say, the governments allocate most of their funding to assist critical government areas, like the health and education sectors. Also, the absence of enough quantity and competitive financial institutions to finance local investments leaves local investors defenseless.

Though many local investors are unable to obtain adequate financing from banking institutions to establish and build wildlife tourism operations, even the eligible ones fail to either seek or finish the repayments due to the exorbitant interest rates from the financial institutions. It is the responsibility of locations or governments to recruit favorable financial facilities that prospective local wildlife tourism investors can access.

Political instability in the wildlife tourism industry leads to the loss of orders and regulations. If political conditions are bad, it could discourage local investment. Political instability has a significant impact on tourist reception, movements, and businesses at the locations. The loss of peace and security amid political turmoil also repels international investors, making wildlife tourism investment impossible.

Additionally, unfavorable investment policies can barely provide enough assistance and empowerment to local investors. Most investment policies promote large-scale wildlife tourism ventures over small ones. In essence, local investors can only handle modest projects because they lack the resources and abilities to handle large enterprises. Furthermore, local investors lack the appropriate expertise and information about wildlife tourism investment potential.

Most investment opportunities are offered in foreign media, embassies, and commercial platforms, where local investors may rarely come across them. The locations should have local investment centers for local investors to learn about wildlife tourism initiatives. Furthermore, bureaucracy and corruption are some of the primary factors deterring local investors from participating in wildlife tourism firms.

In their home nations, local investors are incapable of paying licensing officials to secure investment permits. The loss of public ethics among the authorities authorized to issue business permits may also hinder local investors from venturing into wildlife tourism.

Scrutinization of Service Provisions

It is impossible to achieve sustainable wildlife tourism without the collaborative efforts of all stakeholders in the industry. It is a common misconception that the implementation of strategies for wildlife sustainability is exclusively the responsibility of wildlife attractions and destinations. Consequently, the roles of other wildlife players, such as tour operators and wildlife tourists, consist of the enjoyment of wildlife resources.

According to sustainability principles, all tourism stakeholders must assume responsibility for wildlife conservation and utilization. Consequently, every wildlife tourist in the era of responsible tourism must have a comprehensive understanding of the source and status of each wildlife product or service before purchasing or consuming it. Wildlife tourists can achieve this by researching and identifying any irregularities in wildlife tourism products and services.

Wildlife tourism can rely on the ratings and certification schemes provided by internationally trusted organizations to understand the status of wildlife tourism service providers. Nevertheless, some service providers intentionally exaggerate the quality of their products to deceive tourists, which can result in an overreliance on these ratings. It is crucial to examine the lowest star rating from multiple operators, as it frequently underscores the significant concerns raised by wildlife tourists regarding wildlife issues.

In addition to the research a tourist conducts before visiting an attraction, wildlife tourism can also confirm the authenticity of wildlife products or services during the actual visit. When recommending the product to prospective tourists, friends, and family, wildlife tourists can easily identify abnormal signs or activities at the destination and attraction level. When visiting the animal gardens, it is important to observe things like the appropriateness of shelter, adequate food and water, a resting area, and the sufficient size of the space provided to each animal in the zoo.

Wildlife tourists can easily determine animals' health by observing specific physical symptoms in caged animals, such as perforated skin for skin diseases and visible ribs for starved animals. The same applies in the wilderness, where wildlife may turn into beggars as a sign of animals that are always fed by wildlife tourists.

Furthermore, wildlife tourists may notice that several wildlife tourists surround one animal to observe it when hunting, birthing, migrating, or mating, among other activities. Behavior that may obstruct the animal from practicing its natural call.

Wilderness Experience for Sustainability

Sustainable wildlife tourism encourages experiences that involve the observation of wild animals in their natural environment through wildlife safaris. More than that is associated with the observation of caged or captive animals. It is important to understand that some of the captive or caged animals were obtained through illegal means. Also, some of the captive facilities are uncertified by relevant organizations, and it is also hard for a normal tourist to understand the uncertified animal facilities.

In other words, wildlife tourists need to visit captive facilities certified by internationally recognized animal welfare entities. Moreover, when visiting animals in captivity, wildlife tourists should be involved in activities that prioritize animals' welfare and safety by avoiding unethical and cruel wildlife encounters, such as cage hunting, which denies any hunted animal a natural chance of escaping. The avoidance of unethical activities in the captivities can collectively make a difference by discouraging the owners of captivities and unethical wildlife tourists from involving themselves in unethical wildlife encounters.

Sustainable Wildlife Encounters

Sustainable tourism usually supports responsible conservation efforts as it is managed responsibly and collectively for the benefit of current and future generations. However, conservation responsibilities remain in the hands of the owners of wildlife attractions, wildlife tourists, and other important stakeholders. However, each stakeholder has a separate role to play in attaining the sustainable conservation goals of wildlife destinations.

It is important to understand that when specific procedures to manage wildlife tourists and tour operators are weak or ineffective, the situation regularly devolves into disorder. When law enforcement is inefficient, unethical actors are more likely to break tourism rules or laws, as it is argued that morality begins where the regulations end.

That is to say, voluntary compliance with tourism laws is important for achieving wildlife sustainability goals. Wildlife tourists and tour operators should always prioritize the safety and welfare of wildlife species when visiting protected and unprotected landscapes, either without or with little enforcement from government agencies.

Morality should encourage voluntary compliance with tourism, and conservation should influence wildlife tourists and operators to avoid unethical encounters and activities such as overcrowding around a certain attractive animal, distressing the animals with intensive unnatural noises, cruel training, and separation of social animals.

Supportive Legal and Policy Frameworks

Sustainable wildlife tourism is about compliance with the principles of sustainable development. In that case, the alignment of sustainable development goals with national and international wildlife policies is inevitable. In other words, both tourism laws and policies should reflect sustainability goals to facilitate the efficient implementation of sustainable tourism. Apart from voluntary compliance, legal compliance with sustainability goals creates a sense of responsibility among companies and organizations involved in wildlife tourism because of fear of penalties and punishments imposed by the laws.

Clear policies and practical laws instill responsibility commitments in wildlife tourism stakeholders. Incorporating sustainability practices into the state tourism policies and laws. Having practical policies and laws focusing on tourism sustainability practices may minimize environmental impacts, conserve local natural ecosystems, and preserve the local culture at the destination.

A well-composed legal and policy framework ensures not only the longevity of ecosystems and the viability of destinations' authenticity but it also rightly maximizes tourists' experiences by subjecting them to undisturbed natural environments and cultural resources.

Also, appropriate sustainability policies and legal frameworks prioritize the demands of responsible wildlife tourism. By incorporating and earmarking specific tourism activities, we facilitate awareness among tourism stakeholders about sustainable and responsible wildlife tourism.

Today, many wildlife tourists are more conscious of socially, environmentally, and culturally responsible activities than in the past. By adopting the friendly legal and policy framework, sustainable wildlife tourism and businesses enlarge their customer base and cater to the preferences of responsible wildlife tourists.

Moreover, the relevant framework may lead to cutting down substantial costs and operational efficiency in wildlife tourism service businesses and destination management. The legal and policy framework guides cost savings and operational efficiency for wildlife tourist destinations and businesses. The framework specifies and provides operational and management directives to the destinations and wildlife tourism businesses on energy, waste, economy, and social engagements.

Policy specifications and directives on energy-efficient conservation measures, social responsibilities, waste reduction, and limits on greenhouse gas emissions not only benefit international communities but also help cut down on the operational and management costs of tourism businesses and wildlife tourist destinations. The strategic and legal policy implementation of green practices may lead to innovations and differentiations in wildlife tourism products and services.

The embracement of sustainable green practices through the adoption of friendly legal and policy frameworks has facilitated the innovation and adoption of various green technologies by travel and tourism business enterprises globally. The wildlife tourism business recycles and reuses, redesigns products and services, and improves their service based on the stipulated new environmental, economic, and social standards or trends while complying with the policy and legal frameworks.

Virtual reality is one of the technologies that can immensely reduce the carbon footprints in travel and tourism; virtual wildlife tourism contributes to sustainable green practices by reducing carbon footprints caused by physical traveling behaviors. The technology has unique power and convenience in transporting wildlife tourists to attractive wildlife attractions globally while providing a true sense of physical presence and involvement.

The simulated digital environment created by virtual reality through the use of headsets and other gadgets impersonates wildlife tourists into believing that they are in truly wildlife attractions and destinations. Unlike video watching, virtual reality technology provides an interactive virtual environment, empowering the users to interact with other objects and users.

Among the world-famous virtual tourist destinations is the Great Wall of China, which allows tourists to hike the wall without unnecessary physical effort; this creates a fair hiking tourism opportunity for physically disabled and physically able tourists to access the resources. Wildlife tourists can practice virtual mountaineering through a virtual reality tour by flying across Mount Everest.

This offers a similar view to that provided by the helicopter. Moreover, wildlife tourists can have a virtual tour of Niagara Falls State Park when at home. This is not an exception in Africa, where the Ngorongoro Conservation Area offered virtual wildlife tourism during COVID-19; also, Kenya and South Africa are famous for visual wildlife tourism.

Apart from being a convenient option for wildlife tourists and others who cannot physically travel for any reason, virtual reality contributes substantially to the implementation of sustainability practices. Reduction in physical travel cuts down the greenhouse gas production caused by travelers.

The unnecessary physical intercultural interactions between tourists and locals are reduced, but virtually the interaction may increase while providing maximum cultural interaction and understanding. The destruction of the natural environment caused by tourism activities, including littering, feeding wild animals, chemical usage in hotels, and any other impacts caused by the massification of wildlife tourists, gets reduced.

Similarly, innovation through artificial intelligence has significantly contributed to green sustainability practices, specifically through the reduction of carbon emissions. Optimizing travel equipment through the design of travel paths, fuel efficiency, and even suggestions on necessary maintenance requirements leads to significant reductions in travel and tourism carbon footprints. Also, artificial intelligence offers customized or personalized recommendations for travel and tours while promoting sustainability practices. For example, artificial intelligence avoids recommendations for overcrowded, culturally, and environmentally sensitive areas.

Moreover, it leads to enhancements of the experiences of wildlife tourists while recommending responsible practices, including recommending the wildlife tourists to responsible lodges, travel facilities, attractions, and businesses. Artificial intelligence helps in energy efficiency and conservation, empowering travel and tour operators to manage their resources smartly and therefore comply with sustainable green practices.

With artificial intelligence or data-driven technology, travel and tourism facilities are in the position of tracking their water and energy usage as well as waste production. Artificial intelligence makes sure the resources are reasonably and responsibly used by automatically adjusting water and energy in accommodation facilities, resorts, and travel facilities based on the needs and time. Despite the positive contributions of artificial intelligence to sustainable green practices, artificial intelligence does not come with undesirable challenges; the loss of jobs due to automation is one of those challenges.

Incentivization and Socioeconomic Policies

Creating an enabling environment to encourage or facilitate voluntary adoption of sustainability practices among stakeholders is important. Incentivization provides both extrinsic and intrinsic encouragement by creating an enabling environment that stimulates and facilitates voluntary adoption and compliance with sustainability practices and improves efficiency and consistency in tourism sustainability. Incentivization in sustainable wildlife tourism happens in various contexts and levels. At the organization's or company's level, the wildlife tourism business may motivate employees and customers to comply with sustainability practices.

The same occurs at the international, national, and regional levels, where governments and international organizations may incentivize wildlife tourism businesses and companies to comply with sustainability practices. It is the responsibility of local, international, and regional governments to offer social incentives for wildlife tourism businesses to intensify investment security, minimize operating costs, and afford investment capital in line with sustainability practices. Friendly socioeconomic policies are intangible but technically essential tools for empowering and motivating wildlife tourism stakeholders to engage in sustainable wildlife tourism practices.

Enabling wildlife policies maximizes the engagement of wildlife tourism businesses, destinations, and tourists in environmental, social, economic, and cultural responsibilities. The positive impacts emerging from motivated and empowered sustainable wildlife stakeholders may lead to poverty reduction, the creation of ethical jobs, and increased local supply and demand for tourism services. That is to say, the practicality of sustainable wildlife tourism relies on social, political, and economic incentives that motivate wildlife tourism players to adopt and implement sustainability practices.

It is the responsibility of tourism businesses and organizations to incentivize their employees and wildlife tourists through the creation of conducive working conditions and the provision of fair employment opportunities and wages. Additionally, the inclusion of marginalized groups by considering special groups like people with disabilities and widows in the wildlife tourism business practically motivates responsible wildlife tourists. The direct and indirect involvement of wildlife tourists in environmental, social, economic, and cultural issues before, during, and after their visits incentivizes their participation in sustainability practices.

Since sustainable wildlife tourism means environmental, socioeconomic, and cultural responsibilities are everybody's civic obligation, every wildlife tourism stakeholder, including the tourists and locals, is technically responsible for the implementation of sustainability practices. Since sustainable wildlife tourism is a two-way street, it requires every wildlife tourism stakeholder to be responsible in a specific domain; wildlife tourists should be responsible for their part just like the destination. Wildlife tourism businesses can incentivize their customers and locals through the provision of discounts or special vouchers for those who practice environmentally or culturally responsible activities, such as recycling, following responsible dress codes, and volunteering.

Moreover, any staff from the tourism businesses demonstrating responsible practices, such as saving energy, interacting with communities responsibly, innovating and using green technologies, and recycling waste, should be rewarded and recognized officially. Such rewards usually inspire workers and tourists to behave sustainably and responsibly. Economic incentivization is important as far as the voluntary compliance of wildlife tourism players with sustainability principles. Such incentives may occur as financial and non-financial gains or losses that motivate wildlife tourism to engage in responsible wildlife tourism.

As an ingredient of economic incentive or economic responsibility, fair trade manifests as a hidden hindrance and obstacle to sustainable wildlife tourism. Where is the point of being economically responsible when small-scale local tourism businesses unfairly compete with large-scale foreign businesses? Additionally, where is the point of calling wildlife tourism businesses fair trade when foreign tourism businesses, including hotels and tour companies, import the products, such as vegetables and furniture, that could be supplied by the local companies or local communities from abroad? When the governments create an enabling business environment for foreign investments and disregard the local ones?

Moreover, is it possible to have a fair wildlife trade and businesses when the tourism businesses disregard the locals as their business partners? When all the employees in wildlife tourism enterprises are outsourced from abroad? When the wildlife tourist business has only a foreign investment policy that does not recognize the local ones? When only foreign wildlife tourism investors have tax holidays, waiving and discounting? Apart from the above-described scenarios that highlighted some answered questions in sustainable wildlife tourism about wildlife tourism businesses and trade that may motivate or demotivate wildlife tourism stakeholders.

On the other end, what about the international trade in wildlife resources and products fuelled by legal and illegal wildlife tourism activities? Uncontrolled tourist hunting is one of the domains of wildlife tourism maximising wildlife trafficking and poaching, particularly in developing countries. Frequent and unjustifiable wildlife poaching and trafficking incidents demotivate responsible wildlife tourists from engaging in sustainability practices.

In other words, where is the point of wildlife players engaging in sustainable wildlife tourism when the governments and international wildlife regulators are silent and irresponsible about wildlife trafficking and commercial poaching incidents? Stakeholders can easily? When law enforcement monitors wildlife tourism activities and maximizes the efficiency of wildlife tourist hunting, is it lacking or ineffective?

A serious commitment to stopping wildlife trafficking and commercial poaching is urgently needed, as the trafficking of wildlife and their parts is worth nearly US $20 billion every year. The bloody commercial poaching kills keystone species every day; the UNWTO reports that the world is left with only 400,000 African elephants, and nearly 96 elephants are being killed every day.

Fewer than 30,000 wild rhinoceroses survive in the wilderness globally. International, regional, and national socioeconomic policies and programs that create an enabling environment for fair trade are inevitable. One of the certain ways to incentivize sustainable practices to locals and local businesses is to offer recognition and support.

The governments should facilitate the participation of locals in the wildlife tourism business by formulating friendly financial and non-financial policies and initiating tailor-made tourism programs that address their investment capabilities, such as investment capital, business education, exposure, and business experience. The socioeconomic policies in local investments, tax exemptions, business education, local employment, and local purchasing may address most of the wildlife tourism problems in developing countries, therefore motivating wildlife tourism players to adopt sustainability practices.

Power Dynamics in Sustainable Wildlife Tourism

Power dynamics describe how one party balances or unbalances the relationships with another party or parties. Unbalanced or balanced relationships between two or more individuals, groups, or institutions within a social system occur. Sometimes, a group of individuals, an institution, or an organization distributes formal or informal responsibility to establish who is in charge and what their responsibilities and positions are. In pure engineering terms, power dynamics refer to the production and distribution of power in various ways.

It is an engineering concept employed in the design and optimization of a variety of systems, as well as in guaranteeing the system's effectiveness and performance. Power in relationships and businesses can be equal or unequal. When balanced power dynamics in the social system act as a weighing scale that supports and distributes power. Having a one-sided power dynamic that favors one side over another is toxic and unhealthy. Conflict-resolution discussions, community-centered policies, and honesty can all help establish healthy power dynamics.

When it comes to sustainable wildlife tourism, the idea of power dynamics is crucial. In sustainable wildlife tourism, power dynamics distribute and balance power among tourism stakeholders. All stakeholders in wildlife receive equal power and support in a healthy power dynamic.

Nonetheless, the dynamics of power in the wildlife tourism industry have existed for a very long time; they may even be as old as wildlife tourism. In some colonized nations, there are three distinct periods of power dynamics: precolonial, colonial, and postcolonial. Every era has its power structure. When precolonial or colonial governments imposed restrictions on the use of wildlife, they established biased or unhealthy power dynamics. The postcolonial era seeks healthy power dynamics and fair power distribution.

Domhoff says the power dynamics discussion must address who gains and loses. Who holds the reins of power? Who gains authority? And who has power? Scholars like John French and Bertram Raven have attempted to characterize various types of power dynamics relating to wildlife tourism, drawing on the Domhoff recommendations.

The psychologists pinpointed the following categories of power: referent, formal, reward, informational, expert, coercive, and reward. The majority of them are relevant to wildlife tourism and the key elements that determine sustainable and ethical wildlife tourism.

Coercive power dynamics stem from the threat of punishment to one party in the relationship. It is an unhealthy form of power dynamics where the in-charge party uses threats to frighten the non-complying party. The powerful and most successful investors may use their financial and political influence to intimidate the wildlife tourism stakeholders, pressuring them to acquire and use wildlife resources in a harmful manner. This is a common practice in developing countries where international investors or governments threaten other governments and communities, which may prevent them from grabbing the opportunities to invest in areas with enormous and attractive wildlife resources.

Unethical tourism investors engage in illegal investments in culturally and ecologically sensitive areas, showing a blatant disregard for the traditional and local laws at the wildlife tourists' destinations. Consequently, there is a mushrooming of lodges and other tourism facilities in wildlife corridors, breeding areas, migratory areas, sacred sites, and cultural sites.

As far as sustainable wildlife tourism is concerned, this kind of power dynamic is deadly, as it unfairly damages the natural environment of wildlife and interferes with the sociocultural and economic practices of the host community while discouraging people from participating in conservation. This power dynamic is toxic for wildlife tourism. It leads to the misuse of resources. It hinders the pursuit of sustainability goals.

In wildlife tourism, a person's knowledge or skills may affect power dynamics. Access to certain information may also play a role. Wildlife or tourism managers are often more skilled and knowledgeable. They have power that residents may lack. In a relationship around wildlife areas, experts hold power over locals. Their classical skills and knowledge give them this power. However, locals may have unique knowledge of sociocultural tourism.

Tourism experts do not have this. So, locals are more powerful than the experts. This shows a healthy power balance. Each party earns and holds power. Together, their power may help achieve sustainability goals. Such fruitful power dynamics are common in wildlife areas. They practice community-based tourism or natural resource management. In this kind of power dynamic.

Motivational power dynamics often use motivational or stick approaches. They aim to encourage or discourage certain behaviors in wildlife tourism. Those in power must motivate and demotivate practices in sustainable wildlife tourism. Whether these practices are acceptable or not. The wildlife and tourism managers at the destinations must build positive relationships. They should create a friendly environment for tourists, hotel operators, and locals. This will help both tourism businesses and nature conservation.

The managers may encourage stakeholders to follow sustainability practices. They can do this through awareness campaigns and law enforcement. Wildlife managers and tourism players may promote responsible wildlife tourism. They can do this by rewarding those with the best sustainable practices. This leads us to responsibility schemes and awards. They encourage travel and tourism businesses to be environmentally, socially, and culturally responsible. Wildlife tourism stems from nature conservation.

So, some key wildlife stakeholders are often excluded from its benefits. Their role in cost-sharing and community-based schemes may help wildlife tourism and conservation. This could help promote their efforts in sustainable wildlife tourism. These are the communities that lose lives, loved ones, and property to problem animals. These are harsh conservation costs most humans would dislike.

An individual's official positions in various organizations affect wildlife tourism. The directors and CEOs of various companies hold formal power. So do the ministries and presidents. It is the power that stems from the person's position in the wildlife tourism-related sector. With formal power dynamics, the power holder can tell others what to do. It is also important in decision-making on various aspects of wildlife tourism.

Therefore, power dynamics may affect wildlife tourism's sustainability, both directly and indirectly. Government officials can start reforms for sustainable wildlife tourism. This is the case of the Bhutanese and Singaporean governments. They decided to seriously create and enforce their sustainability policies and practices.

That is the right way to use power to sustain wildlife tourism. Moreover, the formal comes from the policy and laws. They decide on stakeholders' involvement in wildlife tourism, a key aspect of sustainability. Only the minister, director, or president can make those decisions.

Such power dynamics are clear in some countries. There, one person decides everything about sustainable wildlife tourism. A formal power dynamic in the wrong hands is deadly. It harms wildlife tourism. Imagine the many reports of poaching in wildlife tourism spots. The irresponsible holder of formal power may cause most poaching. His power is unquestionable.

This power dynamic lets two parties in wildlife tourism connect. They can achieve mutual goals in win-win situations. Promoting partnerships and collaborations is key to sustainable wildlife tourism. But it is often ignored. Connection power dynamics are healthy. They help marginalised stakeholders access wildlife tourism's benefits. It creates a sense of true relationships between various wildlife tourism players.

For example, there are concerns about enclave tourism in developing countries. All bookings, payments, and food purchases for wildlife tourists are done abroad. There is little or no local buying at the destination. In that situation, stakeholders are poorly linked to wildlife tourism. They could mostly sell vegetables, souvenirs, and their cultural resources there. Wildlife professionals and agencies help connect stakeholders to wildlife tourism opportunities. They do this by addressing power dynamics in the connections.

Global Pandemics and Sustainable Wildlife Tourism

A pandemic is an outbreak of infectious diseases. It affects a wide area, including nations and continents, and a large number of people for a long time. It is an infectious disease that usually crosses the country's political boundary. The world has faced several pandemics. These include the Antonine Plague in 165 A.D., the Cyprian Plague in 250 A.D., leprosy in the eleventh century, and the Black Death in 1350. The Black Death killed about one-third of the world's population. In 1492, the Columbian Exchange killed many people in the Americas. The first cholera pandemic occurred in Russia in 1817. The Russian flu in 1889 claimed 360,000 lives.

The Spanish flu in 1918 killed about 50 million people worldwide. HIV/AIDS has existed since 1981, with no cure found yet. COVID-19 is among the most recent pandemics. Cited pandemics show that most global ones cause massive death and sickness. If we do not control infectious diseases, they can spread quickly and kill many people over a large area.

Governments, global communities, and healthcare organizations worry about any hint of a pandemic. It may threaten human lives and health. Pandemics emerge with surprising speed and often go unnoticed. Often, they involve a new disease-causing agent that has no medical cure. This makes it challenging for unprepared governments to respond to outbreaks. They lack the strategies, equipment, experts, and technology to do so.

Most governments and health organizations collaborated to fight the pandemic. They want to save lives and stop the disease from spreading to other countries. This is despite the infected countries' different locations and political ideologies. The global pandemics need authorities to review their travel policies. They should update or create travel rules.

This will prevent the spread of the disease to uninfected areas. Quarantine is a common strategy. It isolates and restricts those exposed to an infectious disease. If a group returns from a disease-affected area, authorities will quarantine them. Moreover, the governments may enforce a total lockdown. This would require everyone to stay indoors for a time to prevent the new disease from spreading.

Travel and social interactions are vital to socio-cultural, environmental, and economic activities. So, businesses are highly affected by the quarantine and lockdown. Such government enforcement slows people's movement and socio-economic activity. It also increases crime due to unmet basic needs.

There is little literature on the pandemic's impact on wildlife tourism. There is little on how, sustainable wildlife tourism can help post-pandemic. Some studies have reported on the pandemic's impact on tourism. However they focused on a few areas of wildlife tourism. They looked at lost tourism and conservation efforts due to lockdowns.

This book reviews the specific challenges of the global pandemic on sustainable wildlife tourism. Like other sectors, wildlife tourism is similarly affected by global pandemics socially, economically, and environmentally. Understanding the challenges faced by wildlife tourism during pandemics is important because travel and tourism contribute nearly 10.3% to global GDP, which makes its contribution more significant than some of the sectors, including the agricultural sector.

According to Statistica Research Development, tourism also provided more than 330 million formal and informal jobs globally in 2023. The figures show that tourism helps social development. It provides jobs, both direct and indirect, to people of different genders, professions, and ages. In short, tourism provides more employment opportunities than many countries and continents do.

Data from the Bureau of Economics shows that, in 2023, tourism created 330 million jobs. The US created 212 million. Also, a 2024 Statista report said that, in Q1 2023, the EU's employment rate was 199 million. This is nearly half of all travel and tourism workers worldwide. Any pandemic that restricts tourists' movement harms wildlife tourism workers.

This blurs the sustainability of socioeconomic activities. Medical isolation, quarantines, and lockdowns limit transactions between tourists and tourism providers. Unfairly, such restrictions on daily gatherings hurt the poor, women, and elders. They also hurt those in the informal sectors. These include freelance tour guides and vendors of souvenirs and food.

These groups depend on daily, low-income work from tourists. Bans on local and international travel block access to tourism markets. This disrupts wildlife tourism, both domestic and international. However, pandemics may harm other socio-economic sectors. In 2020, the ILO warned that COVID-19 threatened the jobs of nearly half of the world's 3.3 billion workers. The pandemics are killing employees. Their dependents, mostly women, children, and elders, are at risk.

This is more severe in low-income countries, particularly the marginalized and Indigenous populations. Moreover, the pandemic affects the food system at all levels. It exposes many to social inequalities. As international borders close, trade restrictions mount. This prevents businessmen from hiking food prices. It weakens food supply chains and reduces access to basic food services.

The pandemic cut international tourist arrivals. Travel rules, quarantines, and lockdowns caused this. As a result, wildlife tourism lost revenue. Low tourism revenues, especially in poor countries, hurt conservation efforts. They limit the funding of vital conservation work. Reduced frequency and efficiency of law enforcement, research, and community engagement. Many informal workers often lose their jobs and lack benefits. Their access to productive assets is reduced. Also, during pandemics, conservation agencies' law enforcement declines. Poachers may increase illegal activities, like subsistence and commercial poaching and deforestation.

Wildlife may be the main source of food, medicine, and energy during pandemics. Also, illegal wildlife trade may increase near core wildlife areas. In short, pandemics harm wildlife tourism. They disrupt sustainable tourism's key pillars. It affects people's participation in tourism as tourists, service providers, and intermediaries. Pandemics cause sickness and death. They reduce the workforce's size, mobility, and energy. This weakens interactions with the tourism industry. Stakeholders in sustainable wildlife are to blame for this.

Moreover, COVID-19 had mixed effects on the environment. Wildlife thrived in protected areas and unprotected habitats. They faced less competition from humans. The COVID-19 pandemic shut down urban and rural areas. This lets wild animals roam free in protected and unprotected lands. Both land and water species regained confidence. They roamed throughout their habitats without restrictions. In 2020, the BBC reported that wildlife surged in some parks.

Authorities locked tourists in their homes. From the city's docks, I could see some aquatic species. I watched dolphins swimming and jumping. Due to COVID-19, they spotted it in Turkey. A drop in cargo ships, boats, and fishermen staying home caused this. In Haifa, people spotted the boars foraging in the streets. The lack of humans lets them roam freely, even in human areas. During COVID-19, wildlife deaths rose in Banke and Bardia National Parks.

Also, wildlife poaching rose in Chitwan and Shivapuri Nagarjun parks. Wildlife roamed freely due to less human disturbance. However, poachers increased their activity due to a lapse in security. Interestingly, the scavengers and begging animals starved. They relied on humans for food and scraps. Without humans, there would be no more food. Researchers observed this in Europe's pigeons and Indian crows.

They usually depend on waste to forage. In some countries, like Poland, an animal welfare group fed a starving, abandoned bird. Vehicle-wildlife collisions are a major global conservation challenge. Each year, the US records 26,000 injuries and countless wildlife deaths from collisions. The pandemic greatly reduced wildlife-vehicle collisions, especially in densely populated states. The Wildlife Justice Commission said that border closures reduced wildlife trafficking in some Asian countries.

Illegal traders found it hard to access Chinese markets to sell their wildlife. The commission said that due to transport issues, some countries, including Vietnam and Cambodia, have stockpiled illegal wildlife meant for China. Due to the reduced profitability of illegal wildlife trophies, the investigators from the Wildlife Justice Commission witnessed the 22 tons of pangolin scales in Vietnam that could not be exported due to restricted international export during COVID-19.

Post-Pandemic Recovery Strategies and the Role of Domestic Markets

Like other sectors, wildlife tourism was not free from global pandemics. The COVID-19 pandemic exposed its vulnerability. This is true in countries that rely on international travelers for tourism. The Covid-19 pandemic hurt wildlife tourism due to shutdowns and quarantines.

This included domestic tourism. However, domestic tourism was more stable than international travel. This was due to strict restrictions on international travel. The weak tourism market forces countries to rethink their reliance on wildlife tourism. However, this alerts the top wildlife tourist sites. It may shift wildlife tourism's impacts to their limits and needs.

It may also affect conservation efforts. Overreliance on international tourism during pandemics harms wildlife and development. It cuts funding for tourism businesses, weakening conservation efforts. It's time for the conservation and tourism agencies to rethink their marketing. They should boost the hopeful domestic tourism market. This will help sustain tourism during future global crises.

Yet, we must consider the domestic tourism market. It calls for sustainability practices. They should cover the attractions' socioeconomic, cultural, and environmental aspects at the destination level. This book studies whether domestic tourism can help recover from the global pandemics. The COVID-19 pandemic has changed wildlife tourists' travel habits.

They now prefer domestic over international travel. This shows that domestic tourism can completely revive post-pandemic tourism. Despite its potential, we must remember some barriers to tourism. These include the affordability of tourism products and services, a willingness to pay for them, the adoption and use of ICT technologies, economic stability, and reliability, security and safety issues, the availability and convenience of travel services, shutdowns, and quarantine durations.

These factors hinder efforts to jumpstart domestic tourism and rebuild the industry. We can only adopt tourism strategies that promote domestic tourism. They are a recovery and resilience-building strategy. But we must overcome all obstacles to domestic tourism.

Domestic wildlife tourists need money to engage in wildlife tourism. Therefore, the key to boosting domestic tourism is making travel affordable for them. It is vital before adopting a domestic tourism recovery strategy. In developing countries, people see tourism as expensive.

Studies show a direct relationship between tourism and the availability of disposable income. So, first, recover disposable income. Then, plan for domestic tourism's recovery. This is clear in developing countries. Foreigners limit travel and tourism there. Awareness creation and tailoring travel and tourism prices for domestic tourism are important.

Imagine the scenario where a servant earns US$130 and must pay US $230 per night in a tourist hotel. What's the point of being a tourist if my salary can't pay for a hotel for one night? High inflation, unemployment, drought, and climate change are hurting many economies. They make them weak and unsustainable. The weak economy prevents wildlife tourism sites from recovering after COVID-19.

High unemployment and inflation make a domestic tourism recovery strategy hard and unfit. Wildlife tourist destinations can cut costs by adopting new technologies. The COVID-19 quarantine and shutdowns blocked tourists from accessing the attractions.

Wildlife tourists can use ICT tools, like VR and AR, to enjoy the attractions' themes during lockdowns. With ICT technologies, it is possible to create and sell virtual reality packages distantly. Wildlife tourists can see realistic views of wildlife attractions, like mountains, waterfalls, and animal migrations, from their homes. However, before the adoption of ICT, a thoughtful improvement of infrastructure and facilities is necessary.

Imagine the inconvenient situations like internet shutdowns and slowdowns, unavailable and unaffordable ICT tools, lack of human resources, unsupportive policies, and high technology illiteracy rates. Each region's historical, political, and economic context shapes how wildlife tourism operates. Without this regional breakdown, the reader might be left with an incomplete picture of the sector's diversity.

Adoption of Environmentally Friendly Technologies

Sustainable wildlife tourism requires innovation and the adoption of green or clean technologies, which lead to a massive reduction or eradication of waste or greenhouse gases in the wildlife landscape or atmosphere. It encourages the adoption of ecological, economical, and socially friendly technologies for the benefit of tourists, communities, and the environment. That is to say, each tourism element at wildlife destinations is encouraged to adopt green technologies for the attainment of wildlife sustainability goals.

Wildlife-friendly infrastructure, superstructures, and attractions are among the elements requiring the adoption of green technologies. In summary, wildlife infrastructure and superstructure, when planned and constructed thoughtfully and sustainably, may improve not only visitors' experiences but also balance wildlife tourism operations and environmental conservation.

Imagine the impacts of excessive waste reduction and mismanagement, uncontrolled usage of conventional energy, and the construction of hotel establishments in ecologically sensitive areas, among others. By innovating and adopting eco-friendly technologies such as electric cars, wind energy, and efficient energy disposal technologies, sustainable wildlife tourism can balance tourist satisfaction, wildlife conservation, and economic benefits.

Certification Schemes for Sustainable Wildlife Tourism

This book approaches the concept of wildlife tourism from a sustainability perspective. The principles and suggested strategies of sustainability have been clearly described in this book. However, the sustainability indicators or evidence to practically prove whether tourism service providers comply with tourism sustainability or not are still unclear.

As mentioned previously, it is simple for service providers to intentionally or mistakenly declare or label their entities as sustainable when they are not. The Brundtland committee only suggested the principles of tourism sustainability without earmarking the governing bodies to approve or certify the compliance of tourism service providers with the principles.

That raises many questions. One of the most intriguing questions could be: Is there any organization that has been certified as a sustainable tourism entity locally, regionally, or globally? The answer is obviously "yes." This question takes this book to the next important debate.

It involves a discussion about tourism sustainability certification schemes. In short, these are local, national, regional, or global accreditation bodies established to certify an organization's compliance with the principles of sustainable tourism.

Technically, a sustainable tourism certification scheme is the practice of distinguishing trustworthy from counterfeit sustainable tourism products or services. Certification, from the perspective of sustainable tourism, entails the act of certifying that tourism services, products, or entities have passed all sustainability tests and therefore practically comply with all the principles or pillars of sustainability as stipulated by reputable certification schemes.

Certification schemes are required in sensitive products, services, or industries to safeguard the interests and welfare of consumers and global communities, as failure to certify the authenticity and standards of the services may have serious consequences.

The schemes are relevant to sustainable tourism because they play a critical role in promoting ethical practices among tourism stakeholders, which is important in instilling environmental, societal, and cultural responsibilities. Currently, many certification bodies are accredited by international standards organizations.

The ISO numbers refer to environmental reporting, which is a tool to help businesses reduce the environmental implications emanating from their business. However, in travel and tourism, Global Sustainable Tourism Council (GSTC) is responsible for the accreditation of most of the reputable certification bodies.

The Global Sustainable Tourism Council

The Global Sustainable Tourism Council (GSTC) is a fully autonomous and independent international organization registered as a non-profit organization in the United States. It was established with tourism stakeholders in collaboration with UN agencies to guide the development and management of sustainable practices in travel, hospitality, and tourism. It typically establishes and manages global sustainable standards, known as the GSTC Criteria, while also providing an assurance model for global sustainable tourism certification.

GST was formed in 2007 as an alliance of 32 pioneers. The founders of the scheme were the United Nations Environment Programme (UNEP), the United Nations World Tourism Organisation (UNWTO), the Rainforest Alliance, and the United Nations Foundation (UN Foundation). Its membership base includes the private sector, government agencies, non-government organizations, hotel tour operators, individuals, and communities.

Being a certified sustainable entity by the GSTC means that the destination, facilities, attraction, community, or individual complies with the minimum stipulated environmental, social, and economic standards recommended by the scheme. The GSTC provides accreditation to certification bodies. To become certified, tourist travel and tourism entities are advised to choose a relevant certification body from the list of accredited certification bodies stipulated on GSTC's website.

Any travel and tourism stakeholder can become certified by completing the following procedures: choosing a certification body, registering or contracting with the certification body, lodging an application package with the certification body with evidence, conducting on-site auditing or physical verification, and lastly, issuing certification.

Being certified implies that the entities could freely and proudly display information about sustainable certification and inform their customers that they comply with the highest environmental, social, and economic stipulations. The certified entities are sometimes called eco-destinations or green destinations. GSTC provides accreditation services for certification bodies to certify destinations, travel, tourism, and hospitality entities. In short, there are many GSTC-accredited certification schemes for sustainable tourism globally.

Spanning more than one country globally, Control Union Certification is one of the schemes that offers a variety of certification programs. Also, United Certification System Limited is an independent international certification body accredited by GSTC for the certification of sustainable entities. The organization is headquartered in the European Union.

The Mauritius Standards Bureau, or the Blue Oasis, exemplifies the accredited certification bodies under GSTC, offering certification services in the areas of hotels, accommodation, and tour operations within Mauritian jurisdiction and territories. Labeled as the Blue Oasis, it is a not-for-profit parastatal organization accredited by the GSTC.

Certification schemes are not free from limitations; other global industries, including food and beverages, have attempted to openly declare and label the ingredients and methods used to make their products, but the producers of sustainably certified tourism products hesitate to declare publicly and transparently whether their products and services are made with socially, environmentally, and economically friendly ingredients or the procedures used to make their products and services. High water consumption, greenhouse gas emissions, the use of ingredients from endangered species, and the utilization of ecologically sensitive areas are some of the facts tourists need to know before consumption.

Inconsistencies, ineffectiveness, and unethical behavior exist among certification schemes, as some schemes would set minimum-based certification standards to attract the intentions of many tourism entities. This results in enormously certified but technically unqualified organizations that mislead tourists.

This sounds like rewarding probable rather than actual adopters of sustainable tourism. Certification schemes should reset their baseline standards to create opportunities for wildlife tourism service providers to address climate change and variability, support the livelihood of natives, help marginalized groups, respect the welfare of wild animals, and create a healthy, responsible economy.

Sustainability Lesson from Bhutanese Government

King Jigme Khesar Namgyel Wangchuck rules Bhutan, a small landlocked kingdom in South Asia's Himalayas. The kingdom was established in 1644 by Shabdrung Ngawang Namgyel. Before 1974, Bhutan was isolated from the rest of the world, with just a few individuals able to enter the kingdom.

Bhutan is a tiny Himalayan kingdom with appealing wildlife-protected areas that are maintained according to sustainable development principles. It is a small Himalayan kingdom blessed with remarkable wildlife, protected environments, and a commitment to sustainable development.

Bhutan's wildlife attractions include Sakteng Wildlife Sanctuary, Bumdeling Wildlife Sanctuary, Jomotsangkha Wildlife Sanctuary, Phibsoo Wildlife Sanctuary, Royal Manas National Park, Jigme Dorji National Park, Jigme Singye Wangchuck National Park, and Phrumsengla National Park.

The network of protected areas in Bhutan covers more than half of the country's surface area. Such wildlife-protected areas are home to a varied range of wildlife species, including the elusive Bengal tiger, red panda, and other avifauna species.

Bhutan's formal engagement in the international tourism business began in 1974, with 287 people visiting the country's many attractions. In 1972, the nation had already implemented numerous regulations controlling tourist businesses. Since the 1970s, Bhutan has taken a unique sustainable strategy based on high value, low volume, and a focus on tourism.

In 2021, the Tourism Council of Bhutan announced the Kingdom of Bhutan's new tourism policy, which promotes Bhutan as a green, sustainable, inclusive, and high-value tourism destination, with a focus on the guiding principle of high-value and low-volume tourism operations for the country's green socioeconomic development.

The Bhutan tourist policy refers to "green" as all tourist operations and activities that have a low or zero carbon footprint, are eco-friendly, and use green energy, which may contribute to environmental conservation. The government invests in green energy to reduce reliance on fossil fuels, which outweighs the conventional technique of assessing the number of visitors to an attraction at a certain moment.

While tourism is responsible for about 11% of global greenhouse gas emissions due to transmission, Bhutan intends to become a carbon-neutral country by 2025. At present, the country is already 72% carbon dioxide negative, absorbing more carbon dioxide gas from the atmosphere than it emits. This makes Bhutan the sole carbon-negative nation in South Asia. Its trees absorb more than nine million tons of CO_2 per year, while its economy, which is structured to reduce fossil fuel usage and waste, emits fewer than four million tons.

Bhutan's tourist industry continues to expand, becoming one of the country's key economic sectors, contributing considerably to socioeconomic development via income and foreign currency generation, as well as job creation. Bhutan had 315,599 tourists in 2019, a 15% increase from the previous year.

Bhutan's tourism strategy defines sustainable tourism by promoting it in a way that takes into account the requirements of current visitors, destinations, and people while also safeguarding and increasing their needs and possibilities in the future. In the case of a high-value location, tourism policy addresses the requirement to provide excellent value for money and income via tourist operations.

It provides an enabling environment by building excellent infrastructure and bringing tourist goods and services to the nation, resulting in high-quality experience tourism. Furthermore, Bhutan's tourism policy defines inclusive tourism as the creation of opportunities for various players, or Bhutanese, to participate equally in the sharing of tourism costs and benefits, including marginalized groups, women, and elders who play important roles in the country's tourism industry.

Bhutan's tourism policy defines the MDPR as the minimum fee paid by all leisure tourists for an all-inclusive package vacation to Bhutan. It comprises both the SDF and payment for the all-inclusive service package (accommodation, meals, guides, and ground transportation inside Bhutan). The current MDPR is US $250 per person per night in March, April, May, September, October, and November, and US $200 per person per night in the other months of the year.

For those times, the SDF is set at US $65 per person per night. The MDPR will not apply to nations given exemptions, which presently include people from Bangladesh, India, and the Maldives. However, for this kind of leisure tourist, SDF will be used. The SDF is based on the need to manage societal, cultural, and environmental challenges.

According to tourism policy, the SDF is a tourism levy on all leisure tourists per person per night that is applicable throughout the year as a contribution to the government's sustainable development initiatives and to compensate for the negative environmental impacts of tourism. The SDF is imposed in accordance with the Tourism Levy Act of Bhutan 2020.

However, the charge has been altered throughout the years, and tourists who stay for longer periods of time may be eligible for a reduction. This amount of money is being used to support conservation and community development initiatives. In the face of climate change, the nation reduced the daily sustainable development feed to $100 USD to create a compromise between supporting the local economy and employment and safeguarding the natural environment.

The nation follows the high-value, low-volume tourism approach, which generates a significant amount of revenue for infrastructure upgrades and the preservation of natural and cultural heritage. The taxation strategy is not new to Bhutan; several nations, including New Zealand, have used it for many years.

However, what distinguishes Bhutan is its efforts to ensure that all tax revenue is used toward conservation and community development. While many locations and attractions have implemented sustainable tourism initiatives via taxation, a few have followed Bhutan's lead in guaranteeing that the collected tax goes back into conservation and sustainability activities.

Every tourist to Bhutan must pay a Sustainable Development Fee and participate in tourism activities via a tour operator. The nation forbids solo travel and backpacking; thus, all visitors must participate in tourist activities via a tour operator. Tour firms, on the other hand, are adaptable; they offer both group and personal excursions based on the demands of their clients.

Bhutan's Strategies for Tourism Sustainability

This book has illustrated a variety of ways that every wildlife tourist destination may use to accomplish sustainable tourism development. Bhutan, as one of the green destinations, demonstrates the use of sustainability techniques such as pricing and management plans. Bhutan uses tourist pricing and taxing measures such as SDF and MDPR to compensate for the negative environmental consequences of tourism.

The government uses such taxes to provide free healthcare and education to its residents, as well as to offset tourists' carbon footprints by planting trees, cleaning and maintaining trails, and electrifying Bhutan's transportation sector. Furthermore, the taxes are utilized to preserve Bhutan's legacy and pave the road for future generations.

The SDF are expenditures that do not cover direct tourist expenses like lodging, food, transit, and other tourism-related services; they vary from other fees. All government-imposed tourist charges are legally enforced under the Bhutan Tourist Levy Act of 2020.

According to Bhutan's tourist policy, the SDF is enforced by a quota system, with each visitor paying a day charge of US $250 to finance conservation and development initiatives. It is a sustainability charge that ensures tourist activities do not harm Bhutan's cultural and ecologically sensitive regions.

Adopting the notion of tourism carrying capacity for cars and visitors aids in the reduction of traffic congestion and the improvement of visitor experiences. To protect environmental, economic, and cultural integrity, the government restricts the number of cars and visitors by implementing management plans at designated tourist sites and locations of religious, cultural, and natural value.

That means the Bhutanese government is very concerned about cultural, environmental, and safety concerns, such as traffic congestion and vehicular accidents within the attractions, as special guidelines and regulations have been implemented to guide the ownership, driving, and capacity of domestic and foreign tourist vehicles both within and outside of the protected landscapes.

To minimize any negative consequences, the TCB, in partnership with key stakeholders, will analyze the absorptive carrying capacity of selected sites and locations that are susceptible to the detrimental social, environmental, cultural, or other effects of overtourism. With 800,000 citizens, the nation assigns cleaning teams to monitor wildlife regions and mountains and clean up the tracks left by visitors.

Water bottles and any residues are removed by the deployed environmental cleaning teams. This is made feasible by enforcing a quota system in which each tourist must pay a day charge of US $250. It is also known as SDF, which stands for Sustainable Development Fee.

Bhutan's Tourism Council, in conjunction with key stakeholders, works to protect and promote the natural and cultural heritage. As one of its strategies, the Bhutanese government, through the Tourism Council of Bhutan, has developed a mechanism to compensate for the economic opportunities lost by communities living in areas protected and conserved for purposes such as ecotourism by prohibiting development activities. Furthermore, the Tourism Council of Bhutan encourages women, adolescents, and people with disabilities to participate in tourism activities as both visitors and service providers.

Bhutan has four prominent tourist circuits, with the Eastern Circuit experiencing overtourism. As part of its attempt to combat over-tourism, the government has implemented a diversification strategy that has resulted in the development of new tourism goods and clusters. Bhutan has diversified its tourist goods, with a focus on distinctive services and comparative advantage, while also considering the demands for geographical dispersion and year-round tourism.

Medical tourism, including traditional medicine and wellness tourism, nature-based tourism, community-based tourism, village tourism, local cuisine, and sports tourism, are a few examples. However, the introduction of new attractions, tourist goods, and tourism sites takes into account the environmental, economic, and cultural implications for the values, importance, and feelings of local people and locations.

Another technique used by the destination Bhutan is the employment of laws and regulations to ensure the execution of sustainability policies. In such a situation, Bhutan's realistic methods, rules, laws, and standards that guarantee systematic and planned development of sustainable tourist goods make it easier to apply sustainability principles.

Adoption of a quality and standardization plan in which the nation enhances the quality of tourist goods and infrastructure operations by establishing accepted standards for tourism products and services, as well as monitoring and improving existing tourism product offers, packaging, and content creation. This allows the government to encourage domestic tourism by creating goods and services that meet the demands of local visitors.

The need for sustainable and ethical tourism packages is increasing, yet few travellers are ready to pay for sustainable travel. While Bhutan has long been a popular vacation destination for Indians, the imposition of a sustainability charge of US$14.50 has put some travelers off.

As a result, prioritizing socioeconomic, environmental, and cultural integrity has prompted some Indian travelers to either stay here for a while or go to neighboring northeastern Indian states. It is critical to recognize that, while many places throughout the globe see tourist levies as a means of protecting their environment, culture, and economy, such destinations also risk losing visitors looking for economical and cheap vacations.

Unlike many other tourist spots or attractions, Bhutan's conservation strategy encourages citizens to live in wildlife-protected areas. This means that, when properly managed with stakeholder input, tourism has the ability to promote socioeconomic growth, cultural preservation, and biodiversity protection. Tourism is a multisectoral organization that requires ongoing cooperation, communication, and partnership among its stakeholders.

Bhutan's sustainability plans go beyond environmental and economic obligations, incorporating and empowering people. Tour operator firms work with local people to improve their livelihoods and lessen reliance on unethical actions like poaching and deforestation. Tourists stay in green lodgings that comply with all sustainable guidelines. Such facilities are specifically intended to decrease various forms of environmental degradation via methods such as waste reduction, energy efficiency, and water conservation.

In summary, Bhutan is a living example of responsible wildlife tourism, demonstrating to the rest of the world that it is about more than just providing environmental services; it also has economic and communal duties. It is an open chance for many countries to learn from Bhutan at a time when many tourists seek out green and natural areas.

Sustainability Lesson from Singaporean Government

Singapore is a sustainable tourist destination with a tropical climate and plentiful wildlife in natural protected areas. Among other things, the land contains beautiful and melodious birds, animals, a variety of deadly snakes, and the Giant Panda Forest. The country is deeply dedicated to ensuring that wildlife gets world-class welfare and care. By partnering with wildlife conservation partners, the country promotes animal conservation and welfare issues to all visitors visiting wildlife zones.

Singapore, a small island city-state of 734.3 km^2 (283.5 square miles), is the least corrupt in Asia and maybe the fifth least corrupt country in the world, with a crime rate close to zero. According to the National Climate Change Secretariat, Singapore placed 21st out of 155 nations in terms of CO_2 emissions per capita in 2021.

It is one of the nations that are especially sensitive to the impacts of climate change owing to a lack of supporting natural resources, spaces, and climatic conditions essential for large-scale deployment of renewable energy sources. Despite the aforementioned hurdles, the government is still striving to achieve its sustainable green policy.

In 2021, Singapore unveiled the Green Plan 2030, generally known as the Green Plan, as a national vision to steer the country's sustainability campaign. For Singapore, sustainable wildlife tourism, as envisioned in the Green Plan 2030, resembles a five-legged chair, with each leg representing a particular component of visitor sustainability.

The government bears complete responsibility for the economic, environmental, and social pillars of sustainable tourism by dividing them into five dimensions: city in nature, sustainable living, energy reset, green economy, and resilient future. In other words, Singapore accomplishes its aims by adopting the strategies of a city-in-nature, sustainable living, energy reset, green economy, and resilient future.

City in Nature

In collaboration with development partners and conservation stakeholders, the government seeks to build more than 130 hectares of new parks while increasing the quality of existing parks, totaling 170 hectares of natural vegetation. The country aims to plant nearly one million trees throughout Singapore.

Singapore, one of the world's most populated but greenest cities, seeks to grow its protected area network by more than 50%. The OneMillionTrees approach aims to plant over a million trees in 10 years while working with all conservation stakeholders. Singapore walks the talk by maintaining that every household is within a 10-min walk to a park by 2030.

Sustainable Living

This strategy attempts to promote green citizenship by drastically cutting natural resource usage and waste creation by Singaporeans and tourists. Singapore plans to eliminate garbage per capita per day by 20% by 2026, as well as reducing domestic water usage to 130 liters per capita per day.

Furthermore, the government intends to boost green commuting for mass public transit, also known as peak-period modal share. With green commuting, electrifying 50% of the public bus fleet by 2030, and increasing train and bike route networks. Furthermore, the government is expanding its greening efforts to schools by cutting net carbon emissions from the education sector, with 20% of schools reaching zero-carbon emissions.

Green Energy

The goal of the nation's green energy policy is to employ clean, renewable energy sources to reach zero carbon emissions. The deployment of solar energy projects is meeting around 2% of the demand for power. The nation's facilities for treating waste and used water would be entirely energy self-sufficient.

Adoption of green buildings: by 2030, 80% of buildings will use very little clean, green energy. By 2030, the country wants to reduce the amount of energy used in townships that already exist by 15%. A total of 60,000 EV charging stations will be installed nationwide by 2030, and new registration for diesel cars and taxis will end in 2025.

Since *airplanes* are the most widely used form of transportation to and from Singapore, it is important to understand that one of the keys to achieving zero carbon emissions from aviation is using sustainable aviation fuel. Currently, the country runs one of the largest sustainable aviation fuel plants in the world.

In other words, sustainable aviation fuel is used on Singapore Airlines aircraft. Furthermore, by 2025, Changi Airport's new airside light vehicles, forklifts, and tractors are all expected to be electrified by Singapore. By 2040, Singapore wants to make sure that every airside car at Changi Airport runs on cleaner energy.

Resilient Future

As part of its adaptation strategies to address the consequences of climate change, Singapore is scheduled to complete the creation of coastal protection plans for Jurong Island, the North-West Coast, and the City-East Coast. In addition, the government aims to increase the agricultural sector by thirty percent in order to meet Singapore's local nutritional needs in a sustainable manner.

Singapore as a Sustainable Tourism Destination

In February 2023, the Global Sustainable Tourism Council approved Singapore as a sustainable tourist destination. This demonstrates Singapore's efforts and commitment to being a genuinely sustainable tourist destination. The accreditation is based on remarkable examples of cultural sustainability, economic sustainability, social sustainability, and environmental sustainability.

Environmental Sustainability

Singapore's greenery strategy strives to strengthen the ecological linkage between green spaces while simultaneously preserving natural resources sustainably. The OneMillionTrees program highlights the country's determination to attain its greenery target. The country is also making significant progress by exhibiting real-world sustainable methods.

It deepens its commitment to climate change by guaranteeing that more than 40% of buildings are eco-friendly. The government ensures that all tourist facilities are administered in conformity with the principles of sustainability. For instance, Singapore intends to have 60% of its hotel rooms and MICE venues recognized as internationally sustainable by 2025.

Tour operators are increasingly conducting sustainability-focused tours that emphasize Singapore's unique biodiversity, its connection with water, power, waste management, and the local culture. In terms of tourist attractions, each tourism operator must collaborate with the government to meet Singapore's sustainability goals.

Gardens by the Bay is a tourist attraction that has a range of rare and indigenous plant species. Similarly, Sentosa Development Corporation analyses Singapore's environmental policies. The corporation owns many attractions and has invented carbon-neutral drinks.

Social and Cultural Sustainability

It is commonly understood that tourism should not be at the cost of local cultures and communities. For these reasons, Singapore adopts a range of ways to conserve its unique physical and intangible past. Singapore has excellent government rules, legislation, and standards in place to manage and protect its nearly 7000 conserved buildings, places, and monuments. As a consequence, Singapore's cultural legacy is carefully conserved and valued, and tourists are mindful of local customs and traditions.

The Chinatown Past Centre protects and promotes Singapore's cultural past, employing local guides and supporting local businesses. This illustrates social and cultural sustainability in tourism. Furthermore, the government assures that ties between tourists and inhabitants are equal and meaningful. The Singapore Botanic Gardens were named a UNESCO World Heritage Site in 2015. Later in 2020, UNESCO classified Hawker Culture in Singapore as an Intangible Cultural Heritage of Humanity.

Economic Sustainability

Promotion of carbon-efficient investments by establishing Singapore as a sustainable tourism destination, a leading center for green finance and services to help its transition to a low-carbon and sustainable future, Asia's carbon services and trading hub, and the leading regional center for developing new sustainability solutions.

Singapore, as a green tourist destination, knows that economic sustainability in tourism necessitates that the economic advantages be fair and encourage long-term development. As a diversified society, the country makes sure that local communities benefit from tourism. As a result, the government demonstrates its efforts by expanding the economic benefits of tourism activities for citizens.

The benefits include developing stable and long-term employment possibilities by supporting local and resident enterprises, as well as providing appropriate wages and working conditions. It also provides a user-friendly environment and continuing training for locals in travel and tourism. Singapore allows indigenous people to engage in tourism businesses.

Furthermore, the country invests in social infrastructure and community development. The Tourism Sustainability Program of Singapore provides unique support to any sustainable tourism enterprise throughout its sustainability life cycle. Marina Bay Sands displays its dedication to sustainability by installing a large solar panel and adopting an intelligent building management system that optimizes energy usage and conservation.

Conclusion

Wildlife tourism is one of the earliest forms of tourism, and it is a fascinating endeavour to undertake and invest in. It comprises several categories of activities, spanning from gentle to rigorous. It also depicts certain concepts from other forms of tourism, such as cultural and dark tourism, among others.

It takes place in wildlife areas, some of which are either formally or informally protected. It can also take place in unprotected landscapes, which are commonly termed open areas. In wildlife tourism, protection status is not a limit; what matters is the availability and accessibility of sufficient enticing wildlife resources.

Wildlife resources are subjective in wildlife tourism; their definition and scope may vary depending on the destination and conservation policies, among other factors. Wildlife is defined in this book as a free-ranging organism, cageless or barrierless, that includes both vertebrate and invertebrate animals, plants, and the natural environment.

The broad definition of wildlife provides insightful and comprehensive options from which various categories of wildlife tourism activities are technically derived. For example, the natural environment may include wild flora and animals. In such environments, active and inactive forms of wildlife tourism activities are prominent.

In the renowned wildlife tourism destinations, wildlife tourists encounter keystone and endangered species off their reserves, increasing the risk of inbreeding and further heightening the interactions between humans and wildlife. In short, wildlife tourism has social, cultural, economic, and environmental impacts.

Since the future of wildlife tourism relies on the existence of wildlife species and habitats, the conservation and preservation of wildlife resources have become mandatory. Embracing sustainability practices with consideration of socio-economic and environmental aspects remains one of the best alternatives to sustaining wildlife tourism practices.

As a guiding policy, the principles of sustainability endeavor to remind wildlife tourism stakeholders that their consumption of every tourism service or product should respect the environmental, cultural, and economic dimensions of wildlife

A. A. Mamboleo, *Sustainable Wildlife Tourism*,
https://doi.org/10.1007/978-3-031-97795-4

resources and host communities. Similarly, the shifting of wildlife tourists and wildlife operators from conventional to responsible ones is equally essential. The principles make wildlife tour operators and tourists cognisant of their acts.

The understanding of sustainable and responsible tourism is high among tourism players. However, various stakeholders adopt the principles selectively and purposely by only fulfilling one of the three pillars of sustainable development. Such operators think of sustainable wildlife tourism as environmental conservation activities or as only the provision of social infrastructure to needy communities. In many cases, the selection of which pillar to implement is based on the policy and preferences of the wildlife tourism stakeholders.

Also, wildlife tourism still takes place in the traditional way, where game viewing and hunting are still the major products of wildlife tourism, leading to overtourism in some destinations and the failure of repeating wildlife tourists.

References

Abdullah, S. M. M., Othman, M. S., & Yaacob, F. F. (2019). Stakeholders' involvement and perceptions in supporting community-based ecotourism in Min House Camp, Kubang Kerian. *Journal of Tourism, Hospitality and Environment Management*. https://doi.org/10.35631/jthem.416006

Abraham, J. O., & Mumma, M. A. (2021). Elevated wildlife-vehicle collision rates during the COVID-19 pandemic. *Scientific Reports, 11*, 20391. https://doi.org/10.1038/s41598-021-99233-9

Adams, K., Snyder, J., & Crooks, V. A. (2017). Developing an informational tool for ethical engagement in medical tourism. *Philosophy, Ethics, and Humanities in Medicine, 12*, 4. https://doi.org/10.1186/s13010-017-0045-9

Adams, M. P. (2017). The ethics of organ tourism: Role morality and organ transplantation. *The Journal of Medicine and Philosophy, 42*(6), 670–689. https://doi.org/10.1093/jmp/jhx021

Albert, S., Javier, X. L., Grinham, R. A., Church, J. A., Gibbes, B. R., & Woodroffe, C. D. (2016). Interactions between sea-level rise and wave exposure on reef island dynamics in the Solomon Islands. *Environmental Research Letters, 11*, 054011. https://doi.org/10.1088/1748-9326/11/5/054011

Aljazeera. (2022). *Why Kenya's Rift Valley lakes are going through a crisis.* Retrieved December 16, 2022, from https://www.aljazeera.com/news/2020/8/30/why-kenyas-rift-valley-lakes-are-going-through-a-crisis

Aref, F., & Redzuan, M. (2009). Community capacity building for tourism development. *Journal of Human Ecology, 27*, 21–25. https://doi.org/10.1080/09709274.2009.11906187

Ariel, B. (2012). Deterrence and moral persuasion effects on corporate tax compliance: Findings from a randomized controlled trial. *American Society of Criminology, 5*(1), 27–60. https://doi.org/10.1111/j.1745-9125.2011.00256.x

Atere, A., & T. (2020). *UNODC World Wildlife Crime Report 2020: The COVID-19 pandemic has shown that wildlife crime is a threat not only to the environment and biodiversity but also to human health.* UNODC Nigeria Country Office. Retrieved November 3, 2024, from https://www.unodc.org/unodc/en/data-and-analysis/wildlife.html

Auken, I. (2019). *This is what 2030 could look like if we win the war on climate change*

Aziz, S., Abdullah, M., & Niazi, K. (2023). Protecting coastal tourism through developing tourists' environment responsible behaviour. *Journal of Outdoor Recreation and Tourism, 44*, 100698. https://doi.org/10.1016/j.jort.2023.100698

Baltescu, C. (2020). Responsible tourism practices implemented by tourism companies. *Economic Sciences, 13*(62), 49–54. https://doi.org/10.31926/but.es.2020.13.62.2.5

A. A. Mamboleo, *Sustainable Wildlife Tourism*,
https://doi.org/10.1007/978-3-031-97795-4

Baumard, N., Jean-Baptiste, A., & Dan, S. (2013). A mutualistic approach to morality: The evolution of fairness by partner choice. *Behavioral and Brain Sciences, 36*(1), 59–78. https://doi.org/10.1017/S0140525X11002202

Bayih, E., & Singh, A. (2020). Modeling domestic tourism: Motivations, satisfaction and tourist behavioral intentions. *Heliyon, 6*, e04839. https://doi.org/10.1016/j.heliyon.2020.e04839

BBC. (2019). *How Africa will be affected by climate change*. BBC News Services. Retrieved December 21, 2022, from https://www.bbc.com/news/world-africa-50726701

BBC. (2020). *Coronavirus: Wild animals enjoy the freedom of a quieter world*. British Broadcasting Corporation. Retrieved November 3, 2024 from https://www.bbc.com/news/world-52459487

BBC. (2022). *The snows of Kilimanjaro*. Retrieved November 20, 2022, from https://www.bbc.com/storyworks/travel/fit-for-purpose/climate-change-kilimanjaro/

BEC. (2024). *Calculations by the United States Regional Economic Analysist Project*. United States Department of Commerce. Retrieved November 3, 2024 from https://www.unitedstates-reaproject.org/analysis/comparative-trends-analysis/total_employment/tools/0/0/

Bénabou, R., & Tirole, J. (2011). Identity, morals, and taboos: Beliefs as assets. *Quarterly Journal of Economics, 126*, 805–855.

Bertocchi, D., Camatti, N., Giove, S., & Van der Borg, J. (2020). Venice and over-tourism: Simulating sustainable development scenarios through a tourism carrying capacity model. *Sustainability, 12*(2), 512–527. https://doi.org/10.3390/su12020512

Bertzky, A., et al. (2017). European Commission, Joint Research Centre, Science for the AU-EU Partnership: Building Knowledge for Sustainable Development, Publications Office of the European Union, Luxembourg

Bianchi, R. (2018). The political economy of tourism development: A critical review. *Annals of Tourism Research, 70*, 88–102. https://doi.org/10.1016/j.annals.2017.08.005

Blackstock, K. L., White, V., McCrum, G., Scott, A., & Hunter, C. (2008). Measuring responsibility: An appraisal of a Scottish National Park's sustainable tourism indicators. *Journal of Sustainable Tourism, 16*, 276–297. https://doi.org/10.1080/09669580802154090

Bonaccorsi, G., et al. (2020). Economic and social consequences of human mobility restrictions under COVID-19. *Proceedings of the National Academy of Sciences, 117*(27), 15530–15535.

Bradshaw, E. L., & Bateson, P. (2000). Welfare implications of culling Red Deer (Cervus Elaphus). *Animal Welfare, 9*, 3–24.

Brown, A., et al. (2019). Land raising as a solution to sea-level rise: An analysis of coastal flooding on an artificial island in the Maldives. *Journal of Flood Risk Management, 13*(S1), e12567.

Campos-Arceiz, A., & Blake, S. (2011). Megagardeners of the forest – The role of elephants in seed dispersal. *Acta Oecologica, 37*(6), 542–553. https://doi.org/10.1016/j.actao.2011.01.014

CBS News. (2012). *Can hunting endangered animals save the species*? 60 Minutes

Ceballos-Lascurain, H. (2002). *Tourism, ecotourism and protected areas IUCN Section 9*. Assessment, monitoring and management techniques

Chan, J. (2020). *The five: Changes in animal behavior due to global heating*. The Guardian. Retrieved December 22, 2022, from https://www.theguardian.com/science/2020/jan/19/the-five-changes-in-animal-behaviour-due-to-global-heating

Chan, J. K., & Xin, T. (2016). Tour operator perspectives on responsible tourism practices: A case of Kinabalu National Park, Sabah. *International Journal of Culture, Tourism and Hospitality Research, 10*, 121–137. https://doi.org/10.1108/IJCTHR-07-2014-0057

Chang, K., & Wang, K. (2023). Pleasure and restriction: The relationships between community tourism experience value and visitor management. *Journal of Outdoor Recreation and Tourism, 42*, 100613. https://doi.org/10.1016/j.jort.2023.100613

Chase, L. C., Amsden, B., & Phillips, R. G. (2012). Stakeholder engagement in tourism planning and development. In M. Uysal, R. Perdue, & M. Sirgy (Eds.), *Handbook of tourism and quality-of-life research. International handbooks of quality-of-life*. Springer. https://doi.org/10.1007/978-94-007-2288-0_28

Chatterjee, S., & Paul, A. (2023). Role of pollinators in plant reproduction and food security: A concise review. *Journal of Agricultural Sciences, 14*(1), 72–79.

Cheer, J. M., Milano, C., & Novelli, M. (2019). Tourism and community resilience in the Anthropocene: Accentuating temporal over-tourism. *Journal of Sustainable Tourism, 27*, 554–572. https://doi.org/10.1080/09669582.2019.1578363

Chen, Y. Y. B., & Flood, C. M. (2013). Medical tourism's impact on health care equity and access in low- and middle-income countries: Making the case for regulation. *The Journal of Law, Medicine & Ethics, 41*(1), 286–300.

Churchland, P. (2011). *Braintrust: What neuroscience tells us about morality*. Princeton University Press.

Clarke, J., Riemann, L., Vousdoukas, M., & Simpson, N. P. (2022). *Raising sea levels may threaten 70% of Africa' heritage sites by 2050*. The conversation. Retrieved January 21, 2023, from https://theconversation.com/rising-sea-levels-may-threaten-70-of-africas-heritage-sites-by-2050-177385

Cooper, C., Fletcher, J., Fyall, A., Gilbert, D., & Wanhill, S. (2008). *Tourism: Principles and practice* (pp. 221–232). Prentice Hall.

Craigie, I., Grech, A., Pressey, R., Adams, V., Hockings, M., Taylor, M., & Barnes, M. (2014). Terrestrial protected areas of Australia. In *Austral Ark* (pp. 560–581). Cambridge University Press. https://doi.org/10.1017/CBO9781139519960.028

Creel, S., M'soka, J., Dröge, E., Rosenblatt, E., Becker, M. S., Matandiko, W., & Simpamba, T. (2016). Assessing the sustainability of African lion trophy hunting, with recommendations for Policy. *Ecological Applications, 26*(7), 2347–2357. https://doi.org/10.1002/eap.1377

Crespo-Pérez, V., Kazakou, E., Roubik, D. W., & Cardenas, R. E. (2020). The importance of insects on land and in water a tropical view. *Current Opinion in Insect Science, 40*, 31–38. https://doi.org/10.1016/j.cois.2020.05.016

Cressey, E. R. (2012). *The Conservation Genetics of a Newly Recognised Cape Peninsula Endemic: Rose's Mountain Toad (Capensibufo rosei)*. Master of Science thesis, Percy FitzPatrick Institute, University of Cape Town, South Africa.

Cruise, M. (2019). Wildlife conservation and trophy hunting in Africa: Historical perspectives and environmental impacts. *Journal of Environmental History, 24*(3), 45–67.

Dahl, R. A. (1961). *Who governs? Democracy and power in an American City*. Yale University Press.

Dana, B. (2006). *Cheney accidentally shoots fellow hunter*. Retrieved September 21, 2023

Derryberry, E. P., Phillips, J. N., Derryberry, G. E., Blum, M. J., & Luther, D. (2020). Singing in a silent spring: Birds respond to a half-century soundscape reversion during the COVID-19 shutdown. *Science, 370*(6516), 575–579.

Digesser, P. (1992). The fourth face of power. *Journal of Politics, 54*(4), 977–1007.

Ditchkoff, S. S., et al. (1998). *Wounding rates of white-tailed deer with traditional archery equipment*. Southeastern Association of Fish and Wildlife Agencies.

Domhoff, G. W. (2007). C. Wright Mills, Floyd Hunter, and 50 years of power structure research. *Michigan Sociological Review, 21*, 1–54.

Doorne, S. (1998). Power, participation, and perception: An insider's perspective on the politics of the Wellington waterfront redevelopment. *Current Issues in Tourism, 1*(2), 129–166.

Dures, S. G., Carbone, C., Loveridge, A. J., Maude, G., Midlane, N., Aschenborn, O., & Gottelli, D. (2019). A century of decline: Loss of genetic diversity in a southern African lion-conservation stronghold. *Diversity and Distributions, 25*(6), 870–879. https://doi.org/10.1111/ddi.12905

DW. (2022). *Kenya says drought killed over 200 elephants this year*. Retrieved January 23, 2023, from https://www.dw.com/en/kenya-says-drought-killed-over-200-elephants-this-year/a-63650823

EastAfrican. (2021). Babu wa Loliondo, the Tanzanian self-declared miracle healer, dies. The East African Newspaper. Retrieved February 14, 2023, from https://www.theeastafrican.co.ke/tea/news/east-africa/babu-wa-loliondo-tanzanian-self-declared-miracle-healer-dies-3493144

Ellemers, N. (2017). *Morality and the regulation of social behavior: Groups as moral anchors*. Routledge.

Ellemers, N., van der Toorn, J., Paunov, Y., & van Leeuwen, T. (2019). Moralitysychology of analysis review and analysis of empirical studies published from 1940 through 2017. *Personality and Social Psychology Review, 23*(4), 332–366. https://doi.org/10.1177/108886831881175

Esfandiar, K., Pearce, J., Dowling, R., & Goh, E. (2023). The extended theory of planned behaviour model and national parks visitors' pro-environmental binning behaviour: A cross-cultural perspective. *Journal of Outdoor Recreation and Tourism, 42*, 100602. https://doi.org/10.1016/j.jort.2022.100602

Figueres, C., & Revett-Carnac, T. (2020). *What the world will look like in 2050 if we don't cut carbon emission in half.* Retrieved March 23, 2024, from https://time.com/5824295/climate-change-future-possibilities/

Fitzsimons, J., Picone, A., Partridge, T., & Cornish, M. (2023). *Protecting Australia's nature: Pathways to protecting 30 percent of land by 2030.* The Nature Conservancy, WWF-Australia, the Australian Land Conservation Alliance, and the Pew Charitable Trusts.

FMI. (2022). *Wildlife tourism market snapshot (2022-2032).* Retrieved December 20, 2022, from https://www.futuremarketinsights.com/reports/wildlife-tourism-sector-outlook-and-analysis

Francis, R., Ganapathy, S., Arif, M. K., Sakkariya, K. K., & Mohan, S. R. (2020). Responsible tourism: Kerala perspective. *Journal of Emerging Technologies and Innovative Research, 7*(4), 1786–1789.

Friedel, M. (2017). *Forests as carbon sinks.* American Forests. Retrieved January 22, 2023, from https://www.americanforests.org/article/forests-as-carbon-sinks/

Futagami, M., & Thusitha, K. (2022). Sustainable tourism certification and ecolabels: A systematic review of literature. *Journal of the Faculty of Foreign Studies, 5*, 23.

Giampiccoli, A., & Saayman, M. (2018). Community-based tourism development model and community participation. *African Journal of Hospitality Tourism and Leisure, 7*, 1–27.

Godínez-Álvarez, H. (2004). Pollination and seed dispersal by lizards. *Revista Chilena de Historia Natural, 77*, 569–577.

Goodwin, H. (2014). *Cape Town declaration on responsible tourism.* Cape Town Declaration on Responsible Tourism - Responsible Tourism Partnership. Retrieved April 12, 2024

Gössling, S., Scott, D., & Hall, C. M. (2020). Pandemics, tourism, and global change: A rapid assessment of COVID-19. *Journal of Sustainable Tourism.* https://doi.org/10.1080/09669582.2020.1758708

Han, S., Ramkissoon, H., You, E., & Kim, M. J. (2023). Support of residents for sustainable tourism development in nature-based destinations: Applying theories of social exchange and bottom-up spillover. *Journal of Outdoor Recreation and Tourism, 43*, 100643. https://doi.org/10.1016/j.jort.2023.100643

Harich, K. F., Treydte, A. C., Ogutu, J. O., Roberts, J. E., Savini, C., Bauer, J. M., & Savini, T. (2016). Seed dispersal potential of Asian elephants. *Acta Oecologica, 77*, 144–177. https://doi.org/10.1016/j.actao.2016.10.005

Harrill, R. (2004). Residents' attitudes toward tourism development: A literature review with implications for tourism planning. *Journal of Planning Literature, 18*(3), 251–266.

Hellmann, J., Byers, J. E., Bierwagen, J., & Dukes, J. S. (2008). Five potential consequences of climate change for invasive species. *Conservation Biology, 3*, 534–543. https://doi.org/10.1111/j.1523-1739.2008.00951.x

Higgings-Desbiolles, F. (2022). *Rethinking tourism so the locals actually benefit from hosting visitors.* Retrieved January 7, 2022, from https://theconversation.com/rethinking-tourism-so-the-locals-actually-benefit-from-hosting-visitors-116066

Hodges, J. R., Turner, L., & Kimball, A. M. (2012). *Risks and challenges in medical tourism: understanding the global market for health services.* ABC-CLIO LLC.

Holakhe. (2017). *The ugly truth about wildlife conservation in Kenya.* Retrieved December 16, 2022, from https://www.aljazeera.com/opinions/2017/8/3/the-ugly-truth-about-wildlife-conservation-in-kenya/

Holakhe. (2019). Global climate change concerns for Africa's Lake Victoria. Retrieved December 16, 2022, from https://uh.edu/news-events/stories/2019/november-2019/11142019-lake-victoria-water-study.php

Holloway, G. (2002). *Cloning to revive extinct species.* CNN.com.

Horizon. (2024). Asia pacific wildlife tourism market size & outlook. Retrieved November 19, 2024, from https://www.grandviewresearch.com/horizon/outlook/wildlife-tourism-market/asia

Hosonuma, N., Herol, M., De, V. S., Fries, R. S. D., Brockhaus, M., Verchot, L., Angelsen, A., & Romijn, E. (2012). An assessment of deforestation and forest degradation drivers in developing countries. *Environmental Research Letters, 7*, 044009. https://doi.org/10.1088/1748-9326/7/4/04400

Hudman, L. E., & Jackson, R. H. (2003). *Geography of travel and tourism in England*. Thomson Delmar Learning.

Hyde, K. F., & Harman, S. (2011). Motives for a secular pilgrimage to the Gallipoli battlefields. *Tourism Management, 32*, 1343–1351.

Iannoti, M. (2022). *15 garden layouts that make perfect sense*. Retrieved January 7, 2022, from https://www.thespruce.com/garden-layouts-ideas-4171059

IFAW. (2020). Protecting wildlife in Africa during COVID-19. International Fund for Animal Welfare. Retrieved November 3, 2024 from https://www.ifaw.org/news/faq-rangers-africa-protect-wildlife-covid-19

Illinois Department of Natural Resources. (2013). How the program works. Retrieved July 25, 2013

ILO, FAO, IFAD, & WHO. (2020). *Impact of COVID-19 on people's livelihoods, their health, and our food systems*. World Health Organization. Retrieved November 3, 2024 from https://www.who.it/news/13-10-2020/item/impact-of-covid-19-on-people's-livelihoods-their-health-and-our-food-systems

IUCN. (2021). *Invasive alien species and climate change*. International Union for Conservation of Nature Issues Brief. Retrieved December 22, 2022, from https://www.iucn.org/sites/default/files/2022-04/ias_and_climate_change_issues_brief_2021.pdf

IUCN. (2023). *Effective protected areas*. Retrieved September 21, 2023, from https://www.iucn.org/our-work/topic/effective-protected-areas

Jo, H., Ikei, H., & Miyazaki, Y. (2023). Physiological and psychological responses of viewing a waterfall image: Crossover study. *International Journal of Environmental Research and Public Health, 20*(1), 565. https://doi.org/10.3390/9jerph20010565

Joffe-Nelson, N., van Riper, C. J., & Stedman, R. C. (2024). Values and pro-environmental behavior: What is the role of trust? *Journal of Outdoor Recreation and Tourism, 46*, 100745. https://doi.org/10.1016/j.jort.2024.100745

Kangalawe, R., Mwakalila, S., & Msolwa, P. (2011). Climate change impacts, local knowledge and coping strategies in the Great Ruaha River Catchment Area, Tanzania. *Natural Resources, 2*(4), 24027. https://doi.org/10.4236/nr.2011.24027

Kliksberg, B. (2012). *Re-examining the relationship between ethics and economy*. Retrieved April 17, 2024, from https://bbvaopenmnd.com/en/articles/re-examining-between-ethics-and-the-economy/pdf

Koens, K., Postma, A., & Papp, B. (2018). Is over-tourism overused? Understanding the impact of tourism in a city context. *Sustainability, 10*(12), 4383–4398. https://doi.org/10.3390/su10124384

Koju, N. P., Kandel, R. C., Acharya, H. B., Dhakal, B. K., & Bhuju, D. R. (2021). COVID-19 lockdown frees wildlife to roam but increases poaching threats in Nepal. *Ecology and Evolution, 11*(14), 9198–9205. https://doi.org/10.1002/ece3.7778

Layne, D. (2017). Impacts of climate change on tourism in the coastal and marine environments of Caribbean Small Island Developing States (SIDS). *Science Review, 2017*, 174–184.

Lee, Y. C. (2020). Communicating sustainable development: Effects of stakeholder-centric perceived sustainability. *Corporate Social Responsibility and Environmental Management, 27*, 1540–1551. https://doi.org/10.1002/csr.1900

Leiper, N. (1990). Tourist attraction systems. *Annals of Tourism Research, 17*(3), 367–384.

Lindsey, P. A. (2015a). Bushmeat, wildlife-based economies, food security, and conservation. *Insight, 35*(3), 472–482.

Lindsey, P. (2015b). Economic contributions of trophy hunting in sub-Saharan Africa. *Journal of Wildlife Economics, 12*(4), 45–60.

Mamula, N. T., Pantic, S. P., Paunovic, I., & Filipovic, S. (2021). Sustainable travel decision-making of Europeans: Insights from a household survey. *Sustainability, 13*, 1960. https://doi.org/10.3390/su13041960

Manente, M., Minghetti, V., & Mingotto, E. (2014). Responsible tourism and CSR. In S. Idowu & O. Schimidpeter (Eds.), *CSR, sustainability, ethics & governance*. Springer.

Maniatis, A. (2016). Tourism and the 'rule of law'. *African Journal of Hospitality, Tourism and Leisure, 5*(1), 1–13.

Mann, M. (2020). *The community tourism guide: Exciting holidays for responsible travelers*. Earthscan Publications Ltd.

Mason, P. (2016). *Tourism impacts, planning, and management* (3rd ed.). Routledge.

Mayer, I., van Els, D. C., & Bots, P. W. G. (2004). Perspectives on policy analyses: A framework for understanding and design. *International Journal of Technology, Policy and Management, 2004*, 2–20. https://doi.org/10.1504/IJTPM.2004.004819

Medina, L. K. (2005). Ecotourism and certification: Confronting the principles and pragmatics of socially responsible tourism. *Journal of Sustainable Tourism, 13*, 281–295. https://doi.org/10.1080/01434630508668557

Melo, S. R. S., Silva, M. E., Melo, F. V. S., & Vo-Thanh, T. (2022). The practice of (un)sustainable tourism in a National Park: An empirical study focusing on structural elements. *Journal of Outdoor Recreation and Tourism, 39*, 100548. https://doi.org/10.1016/j.jort.2022.100548

Mihalic, T., Mohamadi, S., Abbasi, A., & Dávid, L. D. (2021). Mapping a sustainable and responsible tourism paradigm: A bibliometric and citation network analysis. *Sustainability, 13*, 853.

Milano, C., Cheer, J., & Novelli, M. (Eds.). (2019). *Overtourism: Excesses, discontents and measures in travel & tourism*. CABI.

Ministry of National Development. (2021). *The Singapore Green Plan 2030*. Retrieved October 18, 2024, from https://www.greenplan.gov.sg

Mitchell, R. B. (2007). Compliance theory: Compliance, effectiveness, and behavior change in international environmental law. In J. Brunee, D. Bodansky, & E. Hey (Eds.), *Oxford handbook of international environmental law* (pp. 893–921). Oxford University Press.

Morgan, L. (2012). *Arizona organization protects canned' hunting*

Morrison, A. M., Zhong, L., Deng, B., Coca-Stefaniak, J. A., & Yang, L. (2021). Medical, health and wellness tourism research—A review of the literature (1970–2020) and research agenda. *International Journal of Environmental Research and Public Health, 18*, 10875. https://doi.org/10.3390/ijerph182010875

Murray, C. K. (2017). The lion's share? On the economic benefits of trophy hunting. A report for the Humane Society International, prepared by Economists at Large, Melbourne, Australia. Occasional Paper 25

Musambi, E. (2022). Hundreds of elephants, zebras die as Kenya weathers drought. Retrieved November 5, 2022, from https://www.yahoo.com/hundreds-elephants-zebras-die-kenya-1223313615.html

Mzembe, A. N., Lindgreen, A., Idemudia, U., & Melissen, F. (2020). A club perspective of sustainability certification schemes in the tourism and hospitality industry. *Journal of Sustainable Tourism, 28*, 1332–1350. https://doi.org/10.1080/09669582.2020.1737092

National Climate Change Secretariat. (2022). *Singapore emission profile*. Retrieved October 14, 2024, from https://www.nccs.gov.sg/singapores-climate-action/singapores-climate-targets/singapore-emissions-profile/

National Research Council. (1995). *Science and the endangered species act* (p. 21). National Academy Press.

National Shooting Sports Foundation. (2012). *Firearms-related injury statistics*. Industry intelligence reports

Neuschulz, E. L., Mueller, T., Schleuning, M., & Böhning-Gaese, K. (2016). Pollination and seed dispersal are the most threatened process of plant regeneration. *Scientific Reports, 6*, 29839.

Nichols, E., Spector, S., Louzada, J., Larsen, T., Amezquira, S., & Favila, M. E. (2008). Ecological functions and ecosystem services provided by Scarabaeinae dung beetles. *Biological Conservation, 141*(6), 1461–1474. https://doi.org/10.1016/j.biocon.2008.04.011

Nolan, J. H. (2014). *Branding the recreation, park and tourism products*. National Parks and Recreation Association. Retrieved January 14, 2022, from https://www.nrpa.org/parks-recreation-magazine/2014/march/branding-the-recreation-park-and-tourism-product/

Nyembo, L. O., Larbi, I., Mwabumba, M., Selemani, J. R., Dotse, S., Limantol, A. M., & Bessah, E. (2022). Impact of climate change on groundwater recharge in the lake Manyara catchment, Tanzania. *Scientific African, 15*, 2468–2276. https://doi.org/10.1016/j.sciaf.2021.e01072

O'Hara, P., & Sainato, V. (2019). *Why law enforcement organizations fail: Mapping the organizational fault lines in policing* (3rd ed.). Carolina Academic Press.

Orams, M. B. (2002). Feeding wildlife as a tourism attraction: A review of issues and impacts. *Tourism Management, 23*(3), 281–293. https://doi.org/10.1016/S0261-5177(01)00080-2

Ouko, E., Omondi, S., Mugo, R., Wahome, A., Kasera, K., Nkurunziza, E., Kiema, J., Flores, A., Adams, E. C., Kuraru, S., & Wambua, M. (2020). Modeling invasive plant species in Kenya's Northern Rangelands. *Frontiers in Environmental Science, 8*, 69. https://doi.org/10.3389/fenvs.2020.00069

Outdoor Life. (2019). *Why we suck at recruiting new hunters, why it matters, and how you can fix it*

Özgen Çiğdemli, A. Ö. (2021). Overtourism: Issues, risk and responsibility. In A. Hassan & A. Sharma (Eds.), *Overtourism, technology solutions and decimated destinations*. Springer. https://doi.org/10.1007/978-981-16-2474-2_9

Palfrey, J., Smith, L., Johnson, R., & Lee, A. (2022). Global perspectives on protected area ownership and biodiversity conservation. *Journal of Environmental Management, 301*, 113857. https://doi.org/10.1016/j.jenvman.2022.113857.

Papanicolaou, A. (2011). Authenticity and commodification. In O. Moufakkir & P. Burns (Eds.), *Controversies in tourism* (pp. 41–53). CABI.

Papathanassis, A., Katsios, S., & Dinu, N. R. (Eds.). (2018). *Yellow tourism: Crime and corruption in the holiday sector*. Springer.

Pasape, L. (2022). A review of the role of tour operators towards sustaining ecotourism in Tanzania. *Journal of Hospitality Management and Tourism, 13*(2), 27–37. https://doi.org/10.5897/JHMT2022.0320

Payne, D., & Dimanche, F. (1996). Towards a code of conduct for the tourism industry: An ethics model. *Journal of Business Ethics, 15*, 997–1007.

Pedersen, A. (2002). *Managing tourism at world heritage sites a practical manual for world heritage site managers*. UNESCO World Heritage Centre.

Perera, P. P., Newsome, D., Kusuminda, T., & Walker, O. (2019). Reasons for visitor dissatisfaction with wildlife tourism experiences at highly visited national parks in Sri Lanka. *Journal of Outdoor Recreation and Tourism, 25*, 102–112. https://doi.org/10.1016/j.jort.2018.07.004

Pessot, E., Spoladore, D., Zangiacomi, A., & Sacco, M. (2021). Natural resources in health tourism: A systematic literature review. *Sustainability, 13*, 2661.

Petersson-Forsberg, L. (2014). Swedish spatial planning: A blunt instrument for the protection of outdoor recreation. *Journal of Outdoor Recreation and Tourism, 5-6*, 37–47. https://doi.org/10.1016/j.jort.2014.03.003

Platt, J. R. (2013). *What happens when forest elephants are wiped out in an ecosystem?* Scientific American Arabic. Retrieved January 12, 2023, from https://blogs.scientificamerican.com/extinction-countdown/forest-elephants-wiped-out-ecosystem/

Poole, R. M. (2007). *Hunters: For love of the land*. National Geographic Magazine.

Prado, A., Marolleau, B., Vaissiere, B. E., Barret, M., & Torres-Cortes, G. (2020). Insect pollination: An ecological process involved in the assembly of the seed microbiota. *Scientific Reports, 10*, 3575.

Pugh, T. A. M., Lindeskog, M., Smith, G., & Calle, L. (2019). Role of forest regrowth in global carbon sink dynamics. *Biological Sciences, 116*(10), 4382–4387. https://doi.org/10.1073/pnas.181051211

Renny, D. J. (2002). *Merits and demerits of different methods of culling British wild mammals: A veterinary surgeon's perspective*. British Wild Mammals.

Rosen, J. (2012). *Tanzania: Thousands trek to 'miracle healer' who promises HIV cure*. The World. Retrieved February 11, 2023, from https://theworld.org/stories/2012-01-31/tanzania-thousands-trek-miracle-healer-who-promises-hiv-cure

Rosina, K. F., & Frank, L. K. O. (2022). Provide our basic needs or we go out: The COVID-19 pandemic lockdown, inequality, and social policy in Ghana. *Policy and Society, 41*(2), 217–230. https://doi.org/10.1093/polsoc/puac008

Rutz, C., et al. (2020). COVID-19 lockdown allows researchers to quantify the effects of human activity on wildlife. *National Ecology Evolution, 4*(9), 1156–1159.

Saarinen, J. (2021). Is being responsible sustainable in tourism? Connections and critical differences. *Sustainability, 13*, 6599. https://doi.org/10.3390/su13126599

Scott, D., Gossling, S., & Hall, C. M. (2012). International tourism and climate change. *WIREs Climate Change*. https://doi.org/10.1002/wcc.165

Semcer, J. (2019). The financial impact of trophy hunting in Africa. *African Wildlife Economics Review*, 8(2), 112–125.

Serrano-Bernardo, F. A., Bruzzi, L., Toscano, E. H., & Rosúa-Campos, J. L. (2012). Pollutants and greenhouse gases emissions produced by tourism life cycle: Possible solutions to reduce emissions and to introduce adaptation measures. *Air pollution-A comprehensive Perspective*, 105.

Sharpley, R. (2013). Responsible tourism: Whose responsibility? In A. Holden & D. Fennell (Eds.), *The Routledge handbook of tourism and environment* (pp. 382–391). Routledge.

Shaw, G., & William, M. (2002). *Critical issues in tourism: A geographical perspective* (2nd ed.). Blackwell Publishing.

Skitka, L. J. (2010). The psychology of moral conviction. *Social and Personality Psychology Compass, 4*, 267–281.

Tsaur, S. H., Yen, C. H., & Teng, H. Y. (2018). Tourist–resident conflict: A scale development and empirical study. *Journal of Destination Marketing & Management, 10*, 152–163. https://doi.org/10.1016/j.jdmm.2018.09.002

TUKO. (2022). *Mt Kenya and Mt Kilimanjaro are among the mountains to lose their snow by 2050*. Retrieved November 20, 2022, from https://www.tuko.co.ke/people/481091-mt-kenya-mt-kilimanjaro-mountains-lose-snow-by-2050-report/

U.S. Fish and Wildlife Service. (2012). *2011 National survey of fishing, hunting, and wildlife-associated recreation* (p. 28). GPO.

UM. (2023). *How is your community planning for tourism*. The University of Minnesota Extension Office. Retrieved January 7, 2022, from https://extension.umn.edu/tourism/developing-local-tourism

UN. (2020). *Climate change is an increasing threat to Africa*. United Nations Climate Change. Retrieved December 23, 2022, from https://unfccc.int/news/climate-change-is-an-increasing-threat-to-africa

UN. (2022). *What is climate? How is it different from the weather?* Retrieved December 16, 2022, from https://www.un.org/en/climatechange/what-is-climate-change

UN. (2023). *Causes and effects of climate change*. United Nations. Retrieved January 23, 2023, from https://www.un.org/en/climatechange/science/causes-effects-climate-change

UNDP. (2022). *Environmental sustainability, climate change, and resilience*. Retrieved November 20, 2022, from https://www.undp.org/tanzania/environmental-sustainability-climate-change-and-resilience

UNEP. (2019). *Saving livelihoods in Lake Victoria by downscaling climate date*. UN Environment Programme. Retrieved January 22, 2023, from https://www.unep.org/gan/news/story/saving-livelihoods-lake-victoria-downscaling-climate-data

UNEP-WCMC. (2024). *Protected area profile for Australia from the world database on protected areas*. Retrieved www.protectedplanet.net

UNWTO. (2023). *Global code of ethics for tourism*. World Tourism Organization. Retrieved May 23, 2023, from https://www.unwto.org/global-code-of-ethics-for-tourism

UNWTO. (2024). *Wildlife is our world heritage*. UNWTO/CHIMELONG. Retrieved November 17, 2024.

Urton, J. (2021). *Climate change to fuel the increase in human-wildlife conflict, UW biologist says*. University of Washington. Retrieved December 22, 2022 from https://www.washington.edu/news/2021/07/29/human-wildlife-conflict-climate/

USDA. (2023). *Climate change*. United States Department of Agriculture, National Invasive Species Information Center. Retrieved January 22, 2023, from https://www.invasivespeciesinfo.gov/subject/climate-change

UWTO. (2018). *Overtourism? Understanding and managing urban tourism growth beyond perceptions*. UNWTO.

Van der Meer, E., & Dullemont, H. (2021). Human-carnivore coexistence: Factors influencing stakeholder attitudes towards large carnivores and conservation in Zimbabwe. *Environmental Conservation, 48*(1), 48–57. https://doi.org/10.1017/S0376892920000491

Wanga, J. O., & Mossberg, L. (2014). Stakeholder involvement in tourism destination development: A case of Dunga Beach and Wetland, Kisumu County, Kenya. *Journal of Arts and Humanities, 2*, 13–17.

Whitfield, J. (2003). Sheep horns downsized by hunters' taste for trophies. *Nature, 426*, 595.

Wiens, D., Rourke, J. P., Casper, B. B., Rickart, E. A., LaPine, T. R., & Peterson, J. (1983). Channing: A nonflying mammal pollination of Southern African Proteas. *Annals of the Missouri Botanical Garden, 70*(1), 1–31.

Wildlife Justice Commission. (2020). *New analysis: Measures to combat COVID-19 impact wildlife trafficking, but criminal networks are gearing up to increase operations*. Retrieved https://wildlifejustice.org/new-analysis-measures-to-combat-covid-19-impact-wildlife-trafficking/

World Economic Forum. (2023). Retrieved March 23, 2023, from https://www.weforum.org/agenda/2019/10/what-happens-if-we-beat-climate-change/

WorldAtlas. (2023). Countries with the most protected lands (percentage of area as reserves). Retrieved September 21, 2023, from https://www.worldatlas.com/articles/countries-with-highest-percentage-of-protected-reserve-lands.html

WTTC. (2019). Global wildlife tourism generates five times more revenue than illegal wildlife trade annually. Retrieved December 16, 2022, from https://wttc.org/news-article/global-wildlife-tourism-generates-five-times-more-revenue-than-illegal-wildlife-trade-annually

WWF. (2023). *Wildlife and climate change*. World Wildlife Fund. Retrieved January 22, 2023, from https://www.worldwildlife.org/initiatives/wildlife-and-climate-change

Yeung, J. (2021). *Amazon rainforest is emitting more carbon dioxide than it absorbs in some areas, a study says*. CNN Edition. Retrieved December 22, 2022, from https://edition.cnn.com/2021/07/15/americas/amazon-rainforest-carbon-emission-intl-hnk-scn/index.html

Zenker, S., & Kock, F. (2020). The coronavirus pandemic—a critical discussion of a tourism research agenda. *Tourism Management, 81*, 104164.

Zhang, Q., Zhang, H., & Xu, H. (2021). Health tourism destinations as therapeutic landscapes: Understanding the health perceptions of senior seasonal migrants. *Social Science & Medicine, 279*, 113951. https://doi.org/10.1016/j.socscimed.2021.113951

Zhu, G., Li, X., & Zhang, Y. (2021). Multi-stakeholder involvement mechanism in tourism management for maintaining terraced landscape in important agricultural heritage systems (IAHS) sites: A case study of Dazhai village in Longji Terraces, China. *Landscape, 10*(11), 1146.

MIX
Papier aus verantwortungsvollen Quellen
Paper from responsible sources
FSC® C105338

If you have any concerns about our products,
you can contact us on
ProductSafety@springernature.com

In case Publisher is established outside the EU,
the EU authorized representative is:
Springer Nature Customer Service Center GmbH
Europaplatz 3, 69115 Heidelberg, Germany

Printed by Libri Plureos GmbH
in Hamburg, Germany